Handbook for Heat Exchangers and Tube Banks Design

Donatello Annaratone

Handbook for Heat Exchangers and Tube Banks Design

Springer

Prof. Donatello Annaratone
Via Ceradini 14
20129 Milano
Italy
donatello@annaratone.eu

ISBN 978-3-642-13308-4 e-ISBN 978-3-642-13309-1
DOI 10.1007/978-3-642-13309-1
Springer Heidelberg Dordrecht London New York

Library of Congress Control Number: 2010930772

Cover design: WMXDesign GmbH, Heidelberg

Printed on acid-free paper

Springer is part of Springer Science+Business Media (www.springer.com)

Preface

The recently published book by the author, "Engineering Heat Transfer", already dealt with exact computation of heat exchangers and tube banks. In design computation this is accomplished via corrective factors; the latter makes it possible to compute the actual mean temperature difference by starting from the logarithmic one relative to fluids in parallel flow or counter flow.

As far as verification computation is concerned, corrective factors were introduced to compute a certain characteristic factor correctly, as is fundamental for this type of computation.

Based on the above, the author decided to investigate further, refine, and widen this topic: the outcome of this work has resulted in this handbook.

New types of exchangers were examined; the calculation was refined to produce practically exact values for the factors. The scope of the investigation was increased by widening the range of the starting factors. Furthermore, a greater number of values to be included in the tables was considered. Finally, a few characteristics of certain values of the corrective factors were highlighted.

The first section is an introduction; it summarizes the fundamental criteria of heat transfer and proceeds to illustrate the behavior of fluids in both parallel and counter flow. It also shows how to compute the mean isobaric specific heat for some fluids; it illustrates the significance of design computation and verification computation. In addition, it illustrates how to proceed with heat exchangers and tube banks to carry out both design and verification computation correctly.

Appendix A then includes 36 tables as a reference for design computation, The tables contain the corrective factors required to obtain the actual mean temperature difference by starting from the mean logarithmic temperature difference relative to fluids in parallel flow or counter flow.

Finally, Appendix B includes 35 tables for verification computation. As far as heat exchangers are concerned, it shows the values of factor ψ which is required for this type of computation. The values of the corrective factors for coils and tube banks are also presented.

Milano, Italy Donatello Annaratone

Notation

c = specific heat (J/kgK)

d = diameter (m)

E = efficiency factor

h = enthalpy (kJ/kg)

k = thermal conductivity (W/mK)

M = mass flow rate (kg/s)

m = mass moisture percentage (%)

q = heat per time unit (W)

S = surface (m^2)

t = temperature (°C)

U = overall heat transfer coefficient (W/m^2K)

x = thickness (m)

α = heat transfer coefficient (W/m^2K)

β = characteristic factor

γ = characteristic factor

η = efficiency

φ = corrective factor

χ = corrective factor

ψ = characteristic factor

Δt = temperature difference (°C)

Superscripts

$'$ = heating fluid

$''$ = heated fluid

Subscripts

c = counter flow

e = exchanger

i = inside

l = logarithmic

m = mean

o = outside

p = constant pressure (isobaric), parallel flow

w = wall

1 = inlet (for heating or heated fluid)

2 = outlet (for heating or heated fluid)

Contents

1 Introduction to Computation 1
1.1 General Considerations 1
1.2 Mean Isobaric Specific Heat 3
1.2.1 Water and Superheated Steam 4
1.2.2 Air and Other Gases 4

2 Design Computation 7
2.1 Introduction 7
2.2 Fluids in Parallel Flow or in Counter Flow 8
2.3 The Mean Difference in Temperature in Reality 12
2.3.1 Fluids in Cross Flow 14
2.3.2 Heat Exchangers 15
2.3.3 Coils 19
2.3.4 Tube Banks with Various Passages of the External Fluid . 21

3 Verification Computation 25
3.1 Introduction 25
3.2 Fluids in Parallel Flow or in Counter Flow 25
3.3 Factor ψ in Real Cases 33
3.3.1 Fluids with Cross Flow 33
3.3.2 Heat Exchangers 34
3.3.3 Coils 35
3.3.4 Tube Bank with Various Passages of the External Fluid . 37

Appendix A Corrective Factors for Design Computation 39
A.1 Fluids in Cross Flow 39
A.2 Heat Exchangers 42
A.2.1 Heat Exchangers with 2 Passages of Internal Fluid (Fig. 2.5) 42
A.2.2 Heat Exchangers with 3 Passages of Internal Fluid (Fig. 2.6) 58
A.2.3 Heat Exchangers with 4 Passages of Internal Fluid (Fig. 2.7) 74
A.3 Coils 84

A.3.1 Coils with Fluids in Parallel Flow (Fig. 2.8) 84
A.3.2 Coils with Fluids in Counter Flow (Fig. 2.9) 90
A.4 Tube Banks with Several Passages of External Fluid 100
A.4.1 Tube Banks with Fluids in Parallel Flow (Fig. 2.10) . . . 100
A.4.2 Tube Banks with Fluids in Counter Flow (Fig. 2.11) . . . 104

Appendix B Factor ψ or Corrective Factors for Verification Computation 111
B.1 Fluids in Parallel Flow or Counter Flow 112
B.2 Fluids in Cross Flow 118
B.3 Heat Exchangers 121
B.3.1 Heat Exchangers with Two Passages of Internal Fluid (Fig. 2.5) 121
B.3.2 Heat Exchangers with Three Passages of Internal Fluid (Fig. 2.6) 137
B.3.3 Heat Exchangers with Four Passages of Internal Fluid (Fig. 2.7) 153
B.4 Coils 163
B.4.1 Coils with Fluids in Parallel Flow (Fig. 2.8) 163
B.4.2 Coils with Fluids in Counter Flow (Fig. 2.9) 165
B.5 Tube Banks with Several Passages of External Fluid 170
B.5.1 Tube Banks with Fluids in Parallel Flow (Fig. 2.10) . . . 170
B.5.2 Tube Banks with Fluids in Counter Flow (Fig. 2.11) . . . 173

Chapter 1
Introduction to Computation

1.1 General Considerations

A few preliminary explanations are required before dealing with the main topic.

In what follows all quantities in reference to the heating fluid are characterized by superscript ($'$), whereas those in reference to the heated fluid are characterized by superscript ($''$).

In addition, the inlet temperature into the heat exchanger or in the tube bank of both heating and heated fluid will be characterized by subscript (1), whereas the outlet temperature will be characterized by subscript (2).

As we know, if a heating fluid at temperature t' transfers heat to a heated fluid at temperature t'' the transferred heat by the time unit (expressed in W) is given by

$$q = US\left(t' - t''\right) = US\Delta t \tag{1.1}$$

In (1.1) U is the overall heat transfer coefficient (in W/m^{2}K), S the surface of reference (in m^2) and Δt the difference in temperature between the two fluids (in °C).

Both for heat exchangers and for tube banks the heat transfer occurs through the tube wall. Therefore, the surface of reference can be the either the internal or the external of the tubes.

Both choices are possible provided that the overall heat transfer coefficient is computed with reference to the chosen surface. Of course, the product US is the same in both cases.

As we said, the choice is irrelevant. Nonetheless, to avoid confusion our recommendation is to always adopt the surface licked by the heating fluid. In that case the surface of reference will be the internal one if the heating fluid runs inside the tubes, or the external one if the heating fluid hits the tubes from the outside.

By adopting this criterion the overall heat transfer coefficient in reference to the external surface indicated by U_o is given by

$$U_o = \frac{1}{\dfrac{1}{\alpha'} + \dfrac{x_w}{k}\dfrac{d_o}{d_m} + \dfrac{1}{\alpha''}\dfrac{d_o}{d_i}} \tag{1.2}$$

D. Annaratone, *Handbook for Heat Exchangers and Tube Banks Design*,
DOI 10.1007/978-3-642-13309-1_1, © Springer-Verlag Berlin Heidelberg 2010

In (1.2) $\alpha'$and α'' are the heat transfer coefficients of the heating fluid and the heated fluid (in W/m^{2}K), respectively, x_w is the thickness of the tube wall (in m), k is the thermal conductivity of the material of the tubes (in W/mK), and d_o, d_m, d_i are the external, medium and internal diameters of the tubes (in m).

On the other hand, if the overall heat transfer coefficient is in reference to the internal surface and indicated by U_i, we have:

$$U_i = \frac{1}{\dfrac{1}{\alpha'} + \dfrac{x_w}{k}\dfrac{d_i}{d_m} + \dfrac{1}{\alpha''}\dfrac{d_i}{d_o}} \tag{1.3}$$

The computation criteria of the heat transfer coefficients α' and α'' are discussed in the specialized literature (for instance in "Engineering Heat Transfer" by the author) with reference to different types of fluid and to its physical and thermal characteristics, its temperature, its dynamic characteristics, as well as its geometrical characteristics of the tubes making up the bank.

Up to this point we assumed the temperatures of both fluids to be constant but in both heat exchangers and tube banks the heating fluid transferring heat cools down, whereas the heated fluid receiving it warms up.

In other words, the heat transfer implies the variability of temperatures of both fluids.

This fact leads to a series of consequences to be discussed in the following chapters.

Here are some preliminary considerations.

The variability of the temperatures of the two fluids implies the necessity to identify a mean difference in temperature to allow the correct calculation of the heat transfer.

In other words (1.1) must be substituted by the following equation:

$$q = US\Delta t_m \tag{1.4}$$

In (1.4) Δt_m is, in fact, the mean difference in temperature.

The specific heat of the fluids which is crucial for the amount of cooling of the heating fluid and for the heating of the heated fluid, varies with temperature. It will be necessary to introduce a mean specific heat, and this requires familiarity with the enthalpy of fluids.

The overall heat transfer coefficient to be considered constant, actually varies with temperature, since the heat transfer coefficients of both fluids vary with it. Therefore, it will be necessary to decide to which temperatures to refer the value of the heat transfer coefficients or the overall heat transfer coefficient for a correct computation of the heat transfer.

The way in which the two fluids interact with each other is crucial. There are two classic types of interaction, one with the fluids in parallel flow and one with the fluids in counter flow (Fig. 1.1).

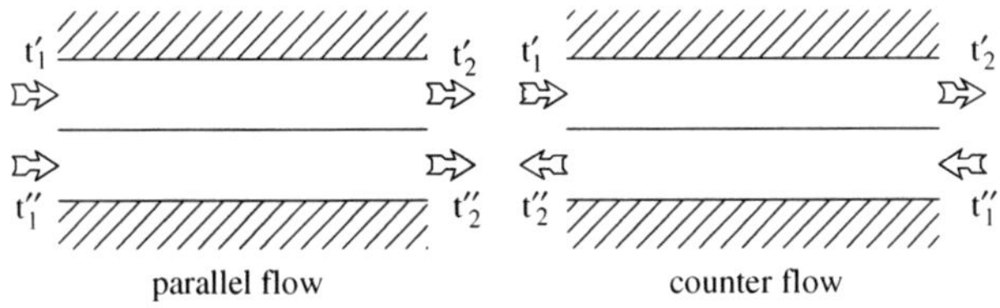

Fig. 1.1

In the first case the heated fluid enters the heat exchanger in the same location of the heating fluid, whereas in the second case the heated fluid enters the heat exchanger where the heating fluid is exiting it.

These situations that simplify the computation of the mean temperature difference will be discussed in Sect. 2.2.

This situation is rare. The path of the two fluids may cross the other one, or it may be a compromise between a path with cross flow and motion in parallel flow or counter flow. This is the case with heat exchangers. Therefore, in all these cases it will be necessary to factor in the actual modality of the heat exchange in ways that will be discussed later on.

We will also point out the possibility for fluids not moving in pure parallel flow or counter flow, but where the heat transfer is such that they can conventionally be considered to be in parallel flow or counter flow. Given the fact, though, that the last assumption is not true, it is necessary to introduce corrective factors.

Finally, there are two types of computation for heat exchangers and tube banks. The first one is the design calculation, consisting of the identification of the exchange surface required to obtain certain results. The second one makes it possible to compute the outlet temperatures of the fluids and the transferred heat, once the exchange surface has been set. This is a verification calculation, and we will discuss both.

1.2 Mean Isobaric Specific Heat

As we shall see, both the design and the verification calculation of the heat exchanger and the tube banks require knowledge of the mean isobaric specific heat of both fluids. Thus, we deem it appropriate to indicate immediately how to proceed in a variety of well-known and less known cases.

The mean isobaric specific heat is given by

$$c_{pm} = \frac{\int_{t_2}^{t_1} c_p dt}{t_1 - t_2}. \tag{1.5}$$

The integral in (1.5) is none other than the difference between enthalpy h_1 corresponding to temperature t_1 and enthalpy h_2 corresponding to temperature

t_2. Considering that the specific heat is usually expressed in J/kgK, whereas the enthalpy is typically expressed in kJ/kg, we have

$$c_{pm} = \frac{h_1 - h_2}{t_1 - t_2} 1000. \tag{1.6}$$

To obtain the required values of c_{pm} it is thus necessary to know the enthalpies of the fluids.

The enthalpy may generally be expressed with an acceptable approximation by the following equation:

$$h = Xt + Yt^2 + Zt^3 \tag{1.7}$$

where t is the temperature of reference of the fluid.

Now we indicate a few equations to be used for the computation of the enthalpy, always expressed in kJ/kg; the temperatures are in °C.

1.2.1 Water and Superheated Steam

The enthalpies for water and superheated steam can be taken exactly from the publication "Properties of Water and Steam in SI-Units – Springer Verlag" or from similar publications.

Yet, for the approximated computation of the enthalpy of water we can adopt the following equation

$$h = 421.96 \frac{t}{100} - 9.36 \left(\frac{t}{100}\right)^2 + 5.74 \left(\frac{t}{100}\right)^3 \tag{1.8}$$

valid for temperatures between 20 and 250°C.

1.2.2 Air and Other Gases

For the enthalpy of air we can adopt the following equation

$$h = 1003.79 \frac{t}{1000} + 37.76 \left(\frac{t}{1000}\right)^2 + 72 \left(\frac{t}{1000}\right)^3 \tag{1.9}$$

valid for $t = 0 - 300$°C.

The following approximated equations are valid, except for flue gas, for temperatures between 0 and 500°C.

Oxygen (O_2)

$$h = 914.2 \frac{t}{1000} + 117.7 \left(\frac{t}{1000}\right)^2 + 22.8 \left(\frac{t}{1000}\right)^3 \tag{1.10}$$

Nitrogen (N_2)

$$h = 1038\frac{t}{1000} + 18.4\left(\frac{t}{1000}\right)^2 + 78.13\left(\frac{t}{1000}\right)^3 \tag{1.11}$$

Carbon dioxide (CO_2)

$$h = 813.3\frac{t}{1000} + 502.3\left(\frac{t}{1000}\right)^2 - 209.5\left(\frac{t}{1000}\right)^3 \tag{1.12}$$

Carbon monoxide (CO)

$$h = 1038.4\frac{t}{1000} + 35.14\left(\frac{t}{1000}\right)^2 + 78.18\left(\frac{t}{1000}\right)^3 \tag{1.13}$$

Methane (CH_4)

$$h = 2149\frac{t}{1000} + 1550.4\left(\frac{t}{1000}\right)^2 + 136.3\left(\frac{t}{1000}\right)^3 \tag{1.14}$$

Flue gas

Based on information in the textbook by the author already mentioned above, the enthalpy of flue may be computed by the following equation:

$$\begin{aligned} h = {} & (971.7 + 10.49m)\frac{t}{1000} + (162.76 - 2.49m)\left(\frac{t}{1000}\right)^2 \\ & - (25.53 - 2.02m)\left(\frac{t}{1000}\right)^3 \end{aligned} \tag{1.15}$$

In (1.15) m is the mass moisture percentage of the gas; (1.15) is valid for $t = 50 - 1200°C$ and for $m = 0 - 12\%$.

Chapter 2
Design Computation

2.1 Introduction

The design computation consists of determining the surface S of the heat exchanger or the tube bank to obtain a certain result.

To that extent, note that for thermal balance we can write that

$$q = M'' c''_{pm} \left(t''_2 - t''_1\right) = \eta_e M' c'_{pm} \left(t'_1 - t'_2\right) \tag{2.1}$$

In (2.1) q is the heat transferred to the heated fluid in the time unit in W, M' and M'' are the mass flow rates of the heating fluid and the heated fluid, respectively, in kg/s, t'_1 and t'_2 are the inlet and outlet temperatures of the heating fluid, t''_1 and t''_2 are the inlet and outlet temperatures of the heated fluid in °C, c'_{pm} and c''_{pm} are the mean isobaric specific heat of both the heating and the heated fluid in J/kgK, and η_e is the actual or assumed efficiency of the heat exchange.

In addition, we know (from Chap. 1) that

$$q = US\Delta t_m. \tag{2.2}$$

For the design computation, once $M', M'', t'_1, t''_1, \eta_e$ are known, we may wish to obtain the exchange of a certain heat q; from (2.1) we obtain the temperatures t'_2 and t''_2, given that the two mean specific heat depend on the four temperatures in question. It is possible instead to impose temperature t'_2 or temperature t''_2 (2.1); still leads to the other unknown temperature and to heat q.

In any case, in the end we have the value of q and the four temperatures.

At this point, if the fluids are in parallel flow or in counter flow we compute the value of Δt_m, corresponding to the mean logarithmic temperature difference, as we shall see later on. If this not the case, we compute the actual mean temperature difference by multiplying the logarithmic one by a corrective factor; in any case we obtain the value of Δt_m.

Once the overall heat transfer coefficient U is computed, we obtain the necessary surface S through (2.2).

As far as the computation of U we indicate which criterion should be followed in our view to compute α' and α''(see Chap. 1)

D. Annaratone, *Handbook for Heat Exchangers and Tube Banks Design*,
DOI 10.1007/978-3-642-13309-1_2,

For the computation of the heat transfer coefficient of the heated fluid it is best to refer to the arithmetic average of both inlet and outlet temperatures, whereas for the computation of the heat transfer coefficient of the heating fluid, it is generally best to refer to the logarithmic average of the two temperatures above, the necessity to refer to film temperature when it is required for the computation of α, notwithstanding.

2.2 Fluids in Parallel Flow or in Counter Flow

If we examine two fluids in parallel flow or in counter flow, the pattern of the temperatures t' and t'' is shown in both Fig. 2.1 and Fig. 2.2.

M' and M'' are the mass flow rates of both fluids, and c'_{pm} and c''_{pm} refer to the mean specific isobaric heat. The overall heat transfer coefficient U is assumed to be constant.

The heat transferred through the elementary surface dS is given by:

$$dq = UdS(t' - t''). \tag{2.3}$$

On the other hand, given that t' decreases with the increase surface and by introducing the exchange efficiency η_e, the same value dq is equal to

$$dq = -\eta_e M' c'_{pm} dt'. \tag{2.4}$$

If the exchange occurs with parallel flow, given that t'' increases with S, from Fig. 2.1 we see that

$$dq = M'' c''_{pm} dt''. \tag{2.5}$$

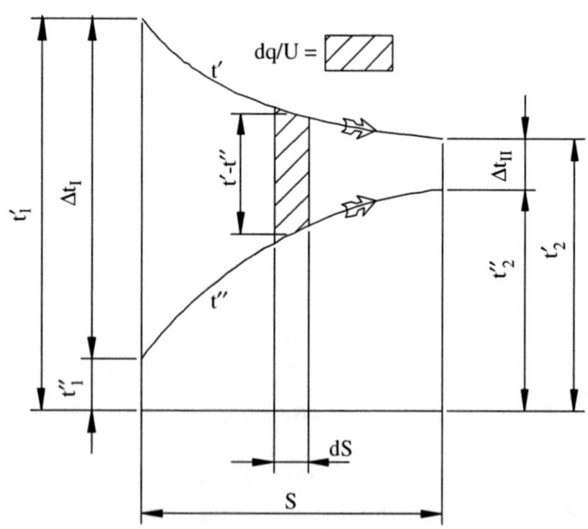

Fig. 2.1 Parallel flow

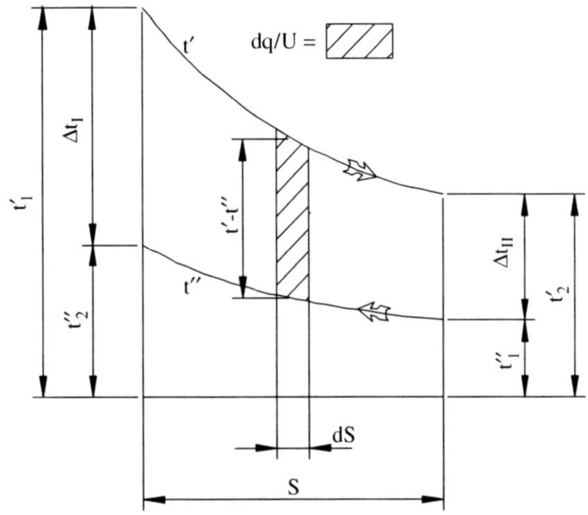

Fig. 2.2 Counter flow

Viceversa, Fig. 2.2 relative to heat transfer during counter flow shows that

$$dq = -M'' c''_{pm} dt''. \tag{2.6}$$

Therefore,

$$d\left(t' - t''\right) = -dq\left(\frac{1}{\eta_e M' c'_{pm}} \pm \frac{1}{M'' c''_{pm}}\right); \tag{2.7}$$

and recalling (2.3)

$$d\left(t' - t''\right) = -UdS\left(t' - t''\right)\left(\frac{1}{\eta_e M' c'_{pm}} \pm \frac{1}{M'' c''_{pm}}\right). \tag{2.8}$$

Here the plus sign indicates parallel flow and the minus sign indicates counter flow.

On the other hand

$$q = M'' c''_{pm}\left(t''_2 - t''_1\right) = \eta_e M' c'_{pm}\left(t'_1 - t'_2\right). \tag{2.9}$$

Thus, with parallel flow

$$\frac{1}{\eta_e M' c'_{pm}} + \frac{1}{M'' c''_{pm}} = \frac{1}{q}\left(t'_1 - t''_1 - t'_2 + t''_2\right), \tag{2.10}$$

and with counter flow

$$\frac{1}{\eta_e M' c'_{pm}} - \frac{1}{M'' c''_{pm}} = \frac{1}{q}\left(t'_1 - t''_2 - t'_2 + t''_1\right) \tag{2.11}$$

The term on the right of the equal sign of both (2.10) and (2.11) (Figs. 2.1 and 2.2) is equal to:

$$\frac{\Delta t_I - \Delta t_{II}}{q}. \tag{2.12}$$

(2.8) can therefore be written as follows:

$$\frac{d\left(t' - t''\right)}{t' - t''} = -\frac{UdS}{q}\left(\Delta t_I - \Delta t_{II}\right); \tag{2.13}$$

and through integration we obtain:

$$\left|-\log_e\left(t' - t''\right)\right|_I^{II} = \frac{US}{q}\left(\Delta t_I - \Delta t_{II}\right); \tag{2.14}$$

then

$$\log_e \frac{\Delta t_I}{\Delta t_{II}} = \frac{US}{q}\left(\Delta t_I - \Delta t_{II}\right). \tag{2.15}$$

Finally,

$$q = US\frac{\Delta t_I - \Delta t_{II}}{\log_e \dfrac{\Delta t_I}{\Delta t_{II}}}. \tag{2.16}$$

The following quantity is the mean logarithmic temperature difference Δt_{ml}:

$$\boxed{\Delta t_{ml} = \frac{\Delta t_I - \Delta t_{II}}{\log_e \dfrac{\Delta t_I}{\Delta t_{II}}}} \tag{2.17}$$

then

$$q = US\Delta t_{ml}. \tag{2.18}$$

The resulting equation is quite similar to (1.1) where instead of the constant difference in temperature between the heating fluid and the heated one, we have the mean logarithmic temperature difference given by (2.17) (of course, U represents U_o and U_i, respectively, depending on whether S is the outside or inside surface of the tubes [see (1.2) and (1.3)].

Another way to proceed is suggested by the fact that, if the ratio $\Delta t_I/\Delta t_{II}$ is not too high, Δt_{ml} does not considerably differ from the mean arithmetic temperature difference equal to:

$$\Delta t = \frac{\Delta t_I + \Delta t_{II}}{2}. \tag{2.19}$$

Therefore, we can write that

$$\Delta t_{ml} = \chi \frac{\Delta t_I + \Delta t_{II}}{2}. \tag{2.20}$$

Based on (2.17) and (2.20), the corrective factor χ is given by

$$\chi = \frac{2\,(\Delta t_I - \Delta t_{II})}{(\Delta t_I + \Delta t_{II}) \log_e \dfrac{\Delta t_I}{\Delta t_{II}}}. \tag{2.21}$$

The value for χ obtained from Fig. 2.3 clearly shows the influence of $\Delta t_I / \Delta t_{II}$ on the reduction of Δt_{ml} with respect to the mean arithmetic temperature difference.

Note that the use of this diagram combined with (2.20) leads to the exact computation of Δt_{ml}.

In the case of fluids in parallel flow, the value of the ratio $\Delta t_I / \Delta t_{II}$ is higher than with fluids in counter flow, thus the value of both χ and Δt_{ml} is smaller. Based on (2.18), it follows that a greater surface with equal transferred heat is needed.

The assumption so far was that the value of U is constant.
In fact, the heat transfer coefficients of both fluids vary with temperature, and so does the value of U. Therefore, it is a question of defining which value of U must be introduced in (2.18).

It is customary to consider the values of the heat transfer coefficients of both fluids corresponding to the average between the inlet and the outlet temperature, and to compute the overall heat transfer coefficient U based on these values of α.

This is the only recommendable (conservative) criterion for heated fluid, even though the behavior of the temperature is not linear. As far as the heating fluid, given the behavior of temperature, it is generally advisable to adopt the

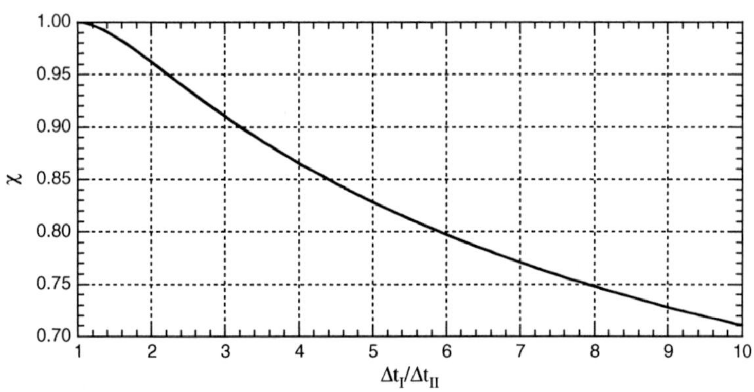

Fig. 2.3

logarithmic average between the inlet and outlet temperatures as reference temperature. Naturally, if the film temperature must be adopted for the computation of the heat transfer coefficient, the temperature of reference must be the average between the temperature mentioned earlier and the wall temperature.

The mean logarithmic temperature of the heating fluid is given by

$$t'_{ml} = \frac{t'_1 - t'_2}{\log_e \dfrac{t'_1}{t'_2}} \tag{2.22}$$

We will come back to this topic when discussing the verification computation.

2.3 The Mean Difference in Temperature in Reality

In real instances the behavior of the fluids, with the exception of fluids with cross flow which are a case in itself, is usually close to the behavior of fluids in parallel flow or counter flow. In general, the most logical methodology to obtain the actual value of Δt_m is to refer to the mean logarithmic difference in temperature in parallel flow or counter flow, and to introduce a corrective factor by which to multiply this difference to obtain Δt_m.

To that extent we introduce three dimensionless factors, the same we will use for the verification computation.

They are:

$$\psi = \frac{t'_2 - t''_1}{t'_1 - t''_1}; \tag{2.23}$$

$$\beta = \frac{\eta_e M' c'_{pm}}{M'' c''_{pm}}; \tag{2.24}$$

$$\gamma = \frac{US}{\eta_e M' c'_{pm}}. \tag{2.25}$$

Since this is a design computation, the inlet and outlet temperatures of both fluids are known, and as a result so is the value of ψ.

Moreover, the value of β is also known.

If we consider the fluids in parallel flow, there is precise connection between the three indicated factors. In fact, based on (3.14) factor γ which is indicated by γ_p, is given by

$$\gamma_p = \frac{1}{1+\beta} \log_e \frac{1}{(1+\beta)\psi - \beta}. \tag{2.26}$$

If we consider the fluids in counter flow instead, and if $\beta \neq 1$, based on (3.23) factor γ indicated with γ_c is given by

$$\gamma_c = \frac{1}{1-\beta} \log_e \left(\frac{1-\beta}{\psi} + \beta \right). \tag{2.27}$$

If $\beta = 1$ instead, from (3.28) we obtain

$$\gamma_c = \frac{1}{\psi} - 1. \tag{2.28}$$

In real instances the value of γ meant to satisfy the imposed value of ψ, is close plus or minus from the value of γ_p or γ_c.

Based on Sects. 2.1 and 2.2, the transferred heat is equal to

$$q = \eta_e M' c'_{pm} \left(t'_1 - t'_2 \right) = US\Delta t_m = \eta_e M' c'_{pm} \gamma \Delta t_m; \tag{2.29}$$

then

$$\Delta t_m = \frac{t'_1 - t'_2}{\gamma}. \tag{2.30}$$

Given that t'_1 and t'_2 are fixed values, we establish that Δt_m is inversely proportional to γ.

If we consider the fluids in parallel flow, instead of (2.30) we must write that

$$\Delta t_{ml(p)} = \frac{t'_1 - t'_2}{\gamma_p} \tag{2.31}$$

where $\Delta t_{ml(p)}$ is the mean logarithmic temperature difference referred to fluids in parallel flow, and γ_p is obtained through (2.26).

Therefore, by introducing the corrective factor χ_p, we may write

$$\boxed{\chi_p = \frac{\Delta t_m}{\Delta t_{ml(p)}} = \frac{\gamma_p}{\gamma}} \tag{2.32}$$

In other words, if the reference is to fluids in parallel flow, after computation of γ_p with (2.26) based on imposed values of ψ and β, the case in question is examined and the real value of γ required to obtain the requested value of ψ is calculated; this way the value of corrective factor χ_p is computed through (2.32).

Thus is possible to compute the value of the actual mean temperature difference Δt_m starting from the value of $\Delta t_{ml(p)}$ relative to the fluids in parallel flow.

The procedure is similar in reference to fluids in counter flow. In that case

$$\boxed{\chi_c = \frac{\Delta t_m}{\Delta t_{ml(c)}} = \frac{\gamma_c}{\gamma}} \tag{2.33}$$

where $\Delta t_{ml(c)}$ is the mean logarithmic temperature difference referred to fluids in counter flow, and γ_c is obtained through (2.27) or (2.28).

Note that with reference to fluids in parallel flow, for the situation to actually be possible we must have

$$\psi > \frac{\beta}{1+\beta}. \tag{2.34}$$

If the reference is to fluids in counter flow instead, and $\beta > 1$, for the situation to actually be possible we must have

$$\psi > \frac{\beta - 1}{\beta}. \tag{2.35}$$

The described process allowed us to build a series of Tables which are included in Appendix A. We refer the reader to this section to make the comparisons discussed in the text.

We did not consider the instances where $\gamma > 6$ since they are unlikely and not advisable. In addition, we neglected those cases where the difference plus or minus between the actual mean temperature difference and the logarithmic one is under 1%, thus to be considered rather insignificant.

In the Tables of Appendix A the missing values to the left of those included correspond to impossible cases or to those where $\gamma > 6$. The missing values to the right of those included correspond to cases where the difference between Δt_m and $\Delta t_{ml(p)}$ or $\Delta t_{ml(c)}$ is less than $\pm 1\%$; for those we can assume the mean logarithmic temperature difference for Δt_m.

2.3.1 Fluids in Cross Flow

The behavior of fluids in cross flow (Fig. 2.4) is closer to that of fluids in counter flow compared to fluids in parallel flow.

So we computed the values of χ_c to include them in the Table A.1.

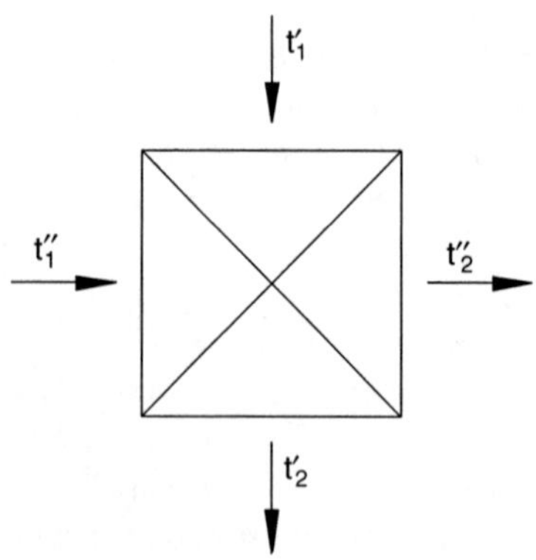

Fig. 2.4 Cross flow

2.3.2 *Heat Exchangers*

2.3.2.1 Heat Exchangers with Two Passages of the Internal Fluid

We consider heat exchangers with two passages of the fluid inside the tubes shown in Fig. 2.5.

As you see, there are four possible combinations indicated by the letters A, B, C and D.

If the number of passages of the fluid external to the tubes is odd, as shown in Fig. 2.5, types A and B which are apparently different from one another, have in fact the same behavior and have the same value of χ.

This depends on the fact that each has one of the two peculiar characteristics of fluids in parallel flow. In fact, in type A the internal fluid enters the tubes in the same location in which the external fluid enters the exchanger; in type B the fluid exits the tubes in the same location in which the external fluid exits the exchanger; this makes their behavior absolutely identical and similar to that of fluids in parallel flow.

If the number of passages of the fluid external to the tubes is even instead, the just described situation occurs for types A and D.

Similar considerations are true for types C and D if we consider an odd number of passages of the external fluid, as described in Fig. 2.5.

Each one has one of the peculiar characteristics of fluid in counter flow. In fact, in type C the internal fluid exits the tubes in the same location in which the external fluid enters the exchanger. In type D the internal fluid enters the tubes in the same location in which the external fluid exits the exchanger. This makes their behavior absolutely identical and similar to that of fluids in counter flow.

If the number of passages of the external fluid is even instead, the just described situation occurs for types B and C.

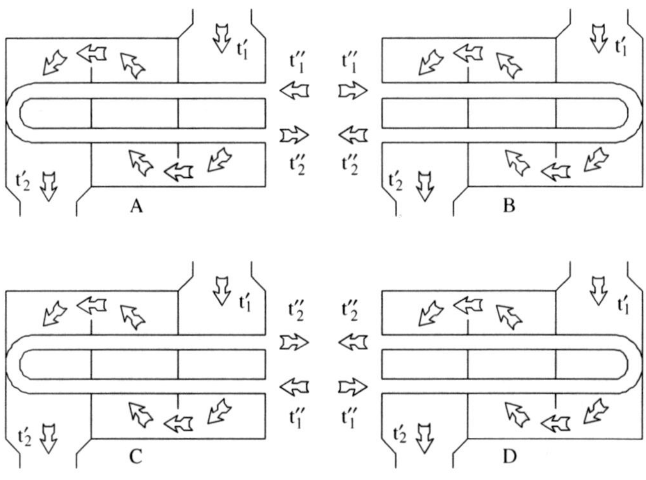

Fig. 2.5 Heat exchangers with two passages of internal fluid

Therefore, for types A and B (or A and D) it would be logical to calculate the value of the corrective factor χ_p, thus referring the requested mean difference in temperature Δt_m to the mean logarithmic difference relative to fluids in parallel flow. Nonetheless, to be able to compare them with types C and D (or B and C) we preferred to compute χ_c; for types C and D (or B and C) the logical solution is undoubtedly that to compute the corrective factor χ_c, thus referring Δt_m to the mean logarithmic difference in temperature relative to fluids in counter flow.

The computation of the values of χ_c is based on a few schemata and assumptions. First of all, the position of the baffles must be such that the exchange surface is divided in equal sections for the various passages of the fluid outside the tubes. Moreover, we assume that the differences in temperature of the different threads of the external fluid annul each other, due to the mixture of the threads occurring with the reversal of the direction of the flow. Thus, the temperature of the external fluid is uniform at the entrance of the new passage.

The analysis was conducted (and this is true for all Tables in Appendix A) by considering β variable between 0.1 and 3.0 and considering ψ variable between 0.04 and 0.96.

The values of χ_c for types A and D or for types B and C with two passages of the external fluid are shown in Tables A.2 and A.3.

The values of χ_c for types A and B or for types C and D with three passages of the external fluid are shown in Tables A.4 and A.5.

The values of χ_c for types A and D or for types B and C with four passages of the external fluid are shown in Tables A.6 and A.7.

Finally, the values of χ_c for types A and B or for types C and D with five passages of the external fluid are shown in Tables A.8 and A.9.

A single passage of the fluid outside the tubes is not considered because in that case the exchanger is reduced to a coil with two sections; its behavior is implied by the section on coils to follow later on.

Analysis of the Tables leads us to interesting considerations.

First of all, it is not surprising that, β and ψ being equal, the value of χ_c and thus of Δt_m for types A and B (or A and D) with reference to Tables A.2, A.4, A.6 and A.8 is always lower than that for types C and D (or B and C) with reference to Tables A.3, A.5, A.7 and A.9.

In addition, the increase in the number of passages of the fluid outside the tubes in types A and B (or A and D) is matched by an increase of Δt_m, whereas in types C and D (or B and C) it decreases. The difference in behavior between types A and D and types B and C which is rather noticeable with 2 passages of the external fluid diminishes with the increase in passages of the external fluid.

Through five passages of the external fluid the values of χ_c relative to the various types of exchangers get considerably closer.

It is rather unlikely that the number of passages of the external fluid is greater than 5; if this should be the case, through, and considering the outcome registered above due to caution we recommend to adopt the values of χ_c included in Table A.8 for all types.

2.3.2.2 Heat Exchangers with Three Passages of the Internal Fluid

If there is just one passage of the external fluid, the exchanger is reduced to a coil with 3 sections. We refer you to the section on coils.

The various types of exchangers with three passages of fluid inside the tubes are shown in Fig. 2.6 and indicated by E, F, G and H.

If the number of passages of the fluid outside the tubes is even, as shown in Fig. 2.6, types E and F have a characteristic in common with the fluids in parallel flow.

In fact, in type E both the internal and the external fluid enter the exchanger in the same position; in type F both fluids exit the exchanger in the same position instead.

Always considering an even number of passages of the external fluid, types G and H have a characteristic in common with the fluids in counter flow.

In fact, if we consider type G we notice that the internal fluid exits from the tubes in the position in which the external fluid enters into the exchanger.

In type H the internal fluid enters the tubes in the position in which the external fluid exits from the exchanger instead.

With an even number of passages of the external fluid type E behaves like type F and type G behaves like type H.

If the number of passages of the external fluid is odd, it is necessary to consider types E and G individually; type E has both characteristics relative to the inlet and outlet of the fluids in common with the fluids in parallel flow; type G has both characteristics relative to the inlet and outlet of the fluids in common with fluids in counter flow.

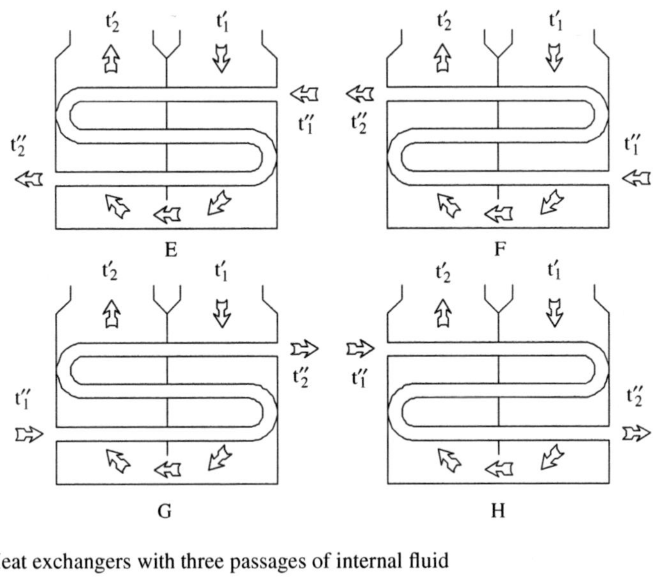

Fig. 2.6 Heat exchangers with three passages of internal fluid

The values of χ_c for types E and F or for types G and H with two passages of the external fluid are shown in Tables A.10 and A.11.

The values of χ_c for type E or for type G with three passages of the external fluid are shown in Tables A.12 and A.13.

The values of χ_c for types E and F or for types G and H with four passages of the external fluid are shown in Tables A.14 and A.15.

Finally, the values of χ_c for type E or for type G with five passages of the external fluid are shown in Tables A.16 and A.17.

The analysis of the values of χ_c in the various Tables shows the following.

For types E and F, moving from 2 to 4 passages of the external fluid (Tables A.10 and A.14), the value of the corrective factor generally slightly decreases.

For types G and H, moving from 2 to 4 passages of the external fluid (Tables A.11 and A.15), the value of the corrective factor generally slightly increases.

The opposite occurs for types E and G with an odd number of passages of the external fluid.

In fact, for type E, moving from 3 to 5 passages of the external fluid (Tables A.12 and A.16) the value of the corrective factor increases; viceversa, for type G, always moving from 3 to 5 passages (Tables A.13 and A.17), the value of the corrective factor decreases.

Finally, note that for type E with an odd number of passages of the external fluid the corrective factor is considerably smaller in comparison with an even number of passages. This is not surprising, given that with an odd number of passages the behavior of the exchanger closely resembles that of fluids in parallel flow.

Similarly, for type G with an odd number of passages of the external fluid the corrective factor is considerably higher compared to an even number of passages; in fact, with an odd number of passages the behavior of the exchanger closely resembles that of fluids in counter flow.

In the unlikely case that the number of passages of the external fluid is greater than 5, given the modest variations taking place after variations in the number of passages, the recommendation is to refer to Tables A.14, A.15, A.16 and A.17, depending on the situation.

2.3.2.3 Heat Exchangers with Four Passages of the Internal Fluid

Now we consider exchangers with 4 passages of the fluid inside the tubes (Fig. 2.7).

If there is just one passage of the external fluid, the exchanger is reduced to a coil with 4 sections. We refer you to the section on coils.

If the number of passages of the fluid outside the tubes is ≥ 3, for some types of exchangers the behavior is quite similar to that of exchangers with 2 passages of the fluid inside the tubes.

Specifically, for types I and L with 3 passages of the external fluid, it is possible to use Table A.4; for types I and N with 4 passages of the external fluid it is possible to use Table A.6; finally, for types I and L with 5 passages of the external fluid it is possible to use Table A.8. The potential errors occurring through this simplification do not exceed 1%. The situation is entirely different, in the case of types I and N

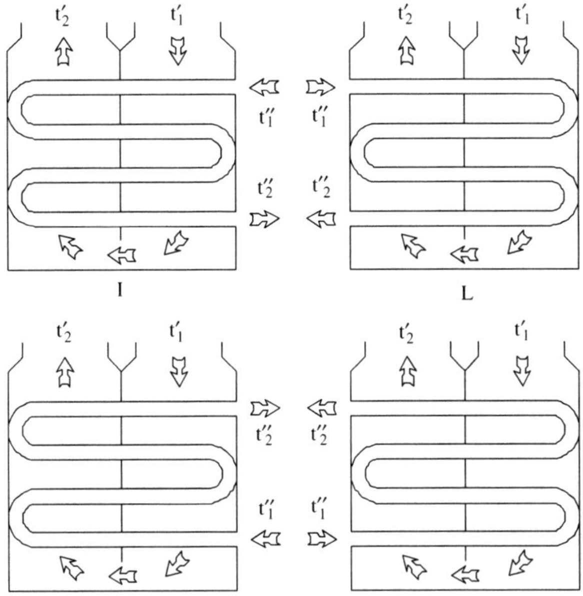

Fig. 2.7 Heat exchangers with four passages of internal fluid

with respect to types A and D, if the exchanger has 2 passages of the external fluid, as shown in Fig. 2.7.

The behavior of the exchanger with 4 passages of the fluid inside the tubes is considerably different from that of an exchanger with 2 passages. Table A.2 cannot be used; for the value of one must refer to Table A.18.

The values of χ_c for types L and M with 2 passages of external fluid are shown in Table A.19.

The values of χ_c for types M and N with 3 passages of the external fluid are shown in Table A.20.

The values of χ_c for types L and M with 4 passages of the external fluid are shown in Table A.21.

Finally, the values of χ_c for types M and N with 5 passages of the external fluid are shown in Table A.22.

As far as the influence of the number of passages of the external fluid on the values of the corrective factor, the same considerations made in reference to the exchangers with 2 passages of the fluid inside the tubes apply. Thus, if the number of passages of the external fluid should be greater than 5, it is recommended to apply caution and refer to Table A.8.

2.3.3 Coils

In the case of coils in Figs. 2.8 and 2.9 it would not be possible to speak of fluids in parallel flow or counter flow. In fact, each section of the coil is hit by the fluid

Fig. 2.8 Coils – parallel flow

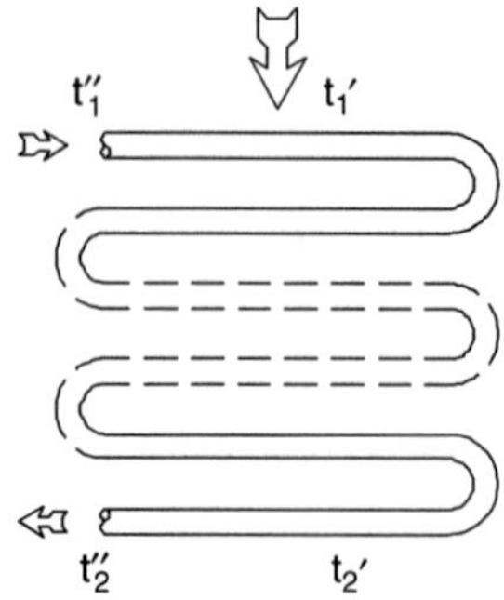

Fig. 2.9 Coils – counter flow

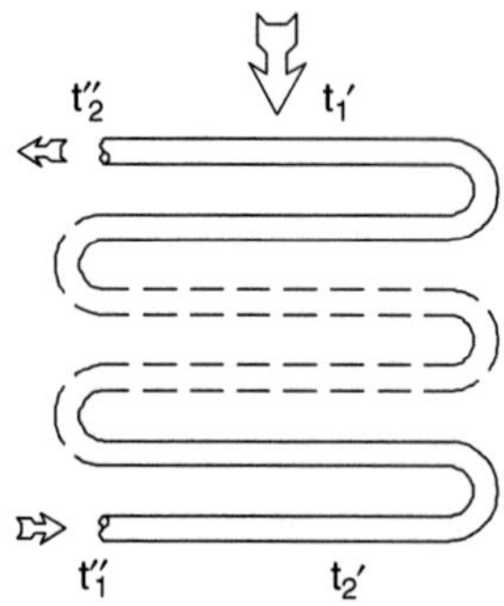

outside the tubes in such a way to be considered cross flow. Therefore, the coil is the sum of elements in which the fluxes are in cross flow.

Usually, though, if the internal fluid enters the coils in correspondence of the inlet in the coil of the external fluid (Fig. 2.8), it is customary to speak of fluids in parallel flow. If the inside fluid enters the coils in correspondence of the outlet of the external fluid (Fig. 2.9), it is customary to speak of fluids in counter flow.

At this point we would like to analyze the topic in-depth both for coils with fluids in parallel flow and those with fluids in counter flow.

2.3.3.1 Coils with Fluids in Parallel Flow

With respect to fluids in real parallel flow they show differences in heat transfer that we would like to highlight. Based on the premises, the corrective factor χ_p is logically calculated.

The considered range is, as for heat exchangers, as follows: $\beta = 0.1 - 3.0$ and $\psi = 0.04 - 0.96$.

Tables A.23, A.24 and A.25 show the values of χ_p relative to coils with 2, 3 and 4 sections, respectively.

We establish that the value of χ_p is always greater than one. This means that the heat transfer occurs with more favourable characteristics compared to those relative to fluids in parallel flow, given that the value of Δt_m is greater than $\Delta t_{ml(p)}$.

As the number of section increases, the value of χ_p gets close to unity. If there are 4 sections there are few instances where $\chi_p > 1.02$; if the number of sections is ≥ 4, giving up the little advantage represented by $\chi_p > 1$, we recommend to adopt the mean logarithmic difference in temperature referred to fluids in parallel flow as value of Δt_m.

2.3.3.2 Coils with Fluids in Counter Flow

Now we consider the coils in Fig.2.9.

Naturally, in this case we calculated the values of χ_c.

The analyzed range is, as usual, as follows: $\beta = 0.1 - 3.0$ and $\psi = 0.04 - 0.96$.

The values of χ_c for a number of sections equal to 2, 3, 4, 6, 8 and 10 are shown in Tables A.26, A.27, A.28, A.29, A.30 and A.31.

As expected, we establish that the values of χ_c are all below unity. This means that the heat transfer is less favorable in comparison with fluids in counter flow, given that Δt_m is smaller than $\Delta t_{ml(c)}$.

The phenomenon is particularly noticeable when the number of sections is small, while it decreases when their number is high.

If the number of sections is $\geq$ 10, the situations where $\chi_c < 0.98$ are rare and unlikely. Therefore, it is possible to conclude that in reality if the number of sections is $\geq$ 10, the coil may be treated as if the fluids were in fact in counter flow by adopting for Δt_m the value of $\Delta t_{ml(c)}$.

In any case, for those situations outlined in Table A.31 where the value of the corrective factor is considerably far from one, it is possible to refer to this Table, even for a number of sections greater than 10.

2.3.4 Tube Banks with Various Passages of the External Fluid

We consider a tube bank consisting of a series of straight tubes; a fluid flows inside the tubes, while another fluid hits the bank outside with a series of passages created through dividing baffles. If there is only one passage of the fluid outside the tubes, these are fluids in cross flow, and we refer the reader to the appropriate section.

The classic device of this type is the recuperative air heater at the end of a steam generator. From now on we will refer to this device but keeping in mind that this type of exchanger can be used even with other fluids, generally gaseous ones.
In air heaters the flue gas is generally located inside the tubes while the air hits the bank outside, but nothing stands in the way of the opposite solution.

The external fluid can enter the heater in correspondence of the inlet to the tubes of the internal fluid, or viceversa with the external fluid entering the heater in correspondence of the exit from the tubes of the internal fluid. Figure 2.10 represents an air heater of the first kind with three passages of the external fluid. Figure 2.11 represents an air heater of the second kind instead.

Clearly, with the first kind the behavior of the fluid through the heater recalls the typical behavior of fluids in parallel flow, whereas the second type is similar to that of fluids in counter flow.

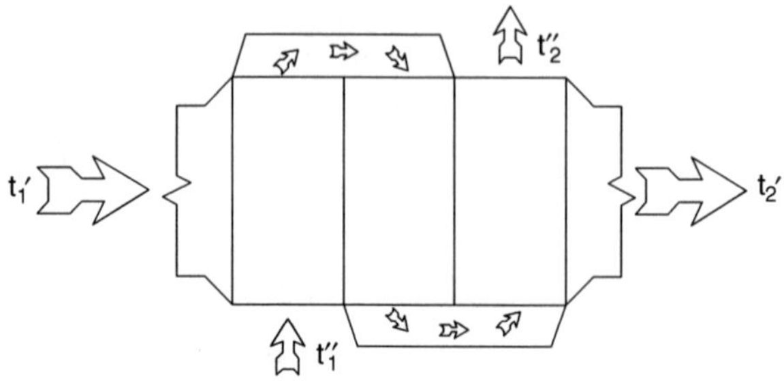

Fig. 2.10 Tube bank with several passages of the external fluid – parallel flow

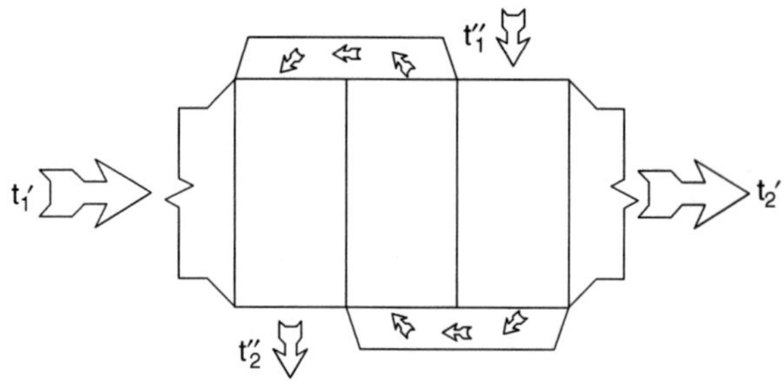

Fig. 2.11 Tube bank with several passages of the external fluid – counterflow

In fact, with these devices it is customary to speak of fluids in parallel flow or in counter flow even this is not exactly true. This topic requires in-depth analysis.

Therefore, we will refer to χ_p for the first type and to χ_c for the second one.

We could consider using the values of χ_p and χ_c already obtained for the coils.

In fact, if the fluid flowing through the tubes of the heater were compared to the fluid hitting the coil, and the fluid hitting the tubes of the heater with the fluid flowing through the coil, the analogy is evident. Still, we must consider that while the temperature of the internal fluid is unique in any position along the coil, the temperature of the fluid hitting the tube bank varies not only depending on the direction of the flux, but also transversally to it. Then the values of the factors cited in relation with the coils are only approximated values.

Another very simple procedure could be as follows. If we assume that the heater is represented by a series of sections where the motion of the fluids is in cross flow, the values of χ_c relative to cross flow can be used for every section, and in the end a global value of χ_p or χ_c is reached to solve the problem. Even this method, though, contains an error. The values of χ_c in Table A.1 are based on uniform temperatures at the inlet of both fluids, while those at the outlet are the average temperatures of the various threads at the exit. In our case we can hypothesize that the temperature of the external fluid is uniform at the inlet of every passage, given the mixture of the threads occurring with the reversal of the flux, but this is certainly not true for the fluid flowing in the tubes. For the latter the division in sections is purely formal because every tube is in one piece where the fluid takes on its own temperature condition which changes from tube to tube.

In view of this, even this method leads to values of χ_p or χ_c yielding only approximated computation.

To obtain more realistic values of χ it is therefore necessary to do a more in-depth analysis to factor in these facts. This is what was done leading to the values in Tables A.32, A.33, A.34, A.35, and A.36.

2.3.4.1 Tube Banks with Fluids in Parallel Flow

We considered the usual range: $\beta = 0.1 - 3.0$ and $\psi = 0.04 - 0.96$.

For the tube bank in parallel flow in Fig. 2.10 the Tables A.32 and A.33 list the values of χ_p for 2 and 3 passages of the external fluid. Of course, they are greater than unity, given that the heat transfer is more favourable than in the case of fluids in parallel flow. We establish that the values of χ_p with 2 passages are greater compared to those with 3 passages. Therefore, the solution with 3 passages is less favourable. Finally, if the passages are ≥ 4 our advice is to give up the modest advantage represented by the fact that in some cases $\chi_p > 1$ by adopting the mean logarithmic temperature difference relative to fluids in parallel flow for Δt_m.

2.3.4.2 Tube Banks with Fluids in Counter Flow

Tables A.34, A.35 and A.36 show the values of χ_c relative to the tube bank in counter flow in Fig. 2.11, respectively, with 2, 3 and 4 passages of the external fluid. The examined range include: $\beta = 0.1 - 3.0$ and $\psi = 0.04 - 0.96$.

Of course, the factors χ_c are below unity since the heat transfer is less favourable compared to the one with fluids in counter flow.

We see that even with 4 passages the difference between Δt_m and the mean logarithmic temperature difference may even be considerable (about up 10%) and it is advisable to take this fact into account.

We did not pursue the investigation any further by examining even solutions with a number of passages greater than 4 given that they are unlikely. In case solutions of this type were adopted, we recommend to conservatively refer to the values of χ_c listed in Table A.36.

Chapter 3
Verification Computation

3.1 Introduction

The verification calculation is in reference to a heat exchanger or a tube bank that were already sized either permanently or temporarily, so that the exchange surface S is known. The verification calculation computes the unknown outlet temperatures of both fluids and the transferred heat.

If the exchanger is sized permanently the verification calculation is done to verify the performance of the exchanger under conditions other than those it was designed for.

The verification calculation can be used even in substitution of the design calculation. This procedure is actually fairly widespread.

In that case the exchange surface is temporary, even though it is possible to compute the heat transfer coefficients of the fluids and the overall heat transfer coefficient. The verification calculation makes it possible to evaluate the performance of the exchanger and to modify its surface if it does not satisfy requirements until the desired result is reached.

The following section will focus on the calculation relative to fluids in parallel flow and counter flow.

These two conditions may be taken as a reference, as we shall see, by introducing corrective factors to obtain the actual condition of the heat exchange. This is the case of coils and tube banks.

In the case of heat exchangers we preferred to compute factor ψ directly instead by including it in the tables; as will be shown later on, this factor is fundamental for the verification computation.

3.2 Fluids in Parallel Flow or in Counter Flow

The symbolism from Sect. 2.2 will be used.

Based on (2.17) and (2.18) and considering fluids in parallel flow, we may write that

D. Annaratone, *Handbook for Heat Exchangers and Tube Banks Design*,
DOI 10.1007/978-3-642-13309-1_3, © Springer-Verlag Berlin Heidelberg 2010

$$q = US\frac{(t'_1 - t''_1) - (t'_2 - t''_2)}{\log_e \frac{t'_1 - t''_1}{t'_2 - t''_2}}. \tag{3.1}$$

Note that

$$\frac{t'_1 - t''_1}{t'_2 - t''_2} = \frac{1 + \frac{t'_2 - t''_1}{t'_1 - t'_2}}{\frac{t'_2 - t''_1}{t'_1 - t'_2} - \frac{t''_2 - t''_1}{t'_1 - t'_2}}. \tag{3.2}$$

Moreover, as in Sect. 2.3 we introduce

$$\psi = \frac{t'_2 - t''_1}{t'_1 - t''_1}; \tag{3.3}$$

$$\beta = \frac{\eta_e M' c'_{pm}}{M'' c''_{pm}}; \tag{3.4}$$

$$\gamma = \frac{US}{\eta_e M' c'_{pm}}; \tag{3.5}$$

On the other hand, the transferred heat is given by:

$$q = M'' c''_{pm} \left(t''_2 - t''_1\right) = \eta_e M' c'_{pm} \left(t'_1 - t'_2\right). \tag{3.6}$$

Then, from (3.4):

$$\beta = \frac{t''_2 - t''_1}{t'_1 - t'_2}. \tag{3.7}$$

We introduce factor ε given by

$$\varepsilon = \frac{t'_2 - t''_1}{t'_1 - t'_2}. \tag{3.8}$$

From (3.3) and (3.7) we obtain:

$$\frac{t'_1 - t''_1}{t'_2 - t''_2} = \frac{1 + \varepsilon}{\varepsilon - \beta}. \tag{3.9}$$

Comparing (3.1) with (3.6) and with reference to (3.5), we have:

$$\gamma = \frac{t'_1 - t'_2}{(t'_1 - t''_1) - (t'_2 - t''_2)} \log_e \frac{t'_1 - t''_1}{t'_2 - t''_2}. \tag{3.10}$$

Recalling (3.9) and (3.4), (3.10) leads to the following:

$$\gamma = \frac{1}{1 + \beta} \log_e \frac{1 + \varepsilon}{\varepsilon - \beta}. \tag{3.11}$$

Then, from (3.11) we obtain:

$$\varepsilon = \frac{1 + \beta e^{(1+\beta)\gamma}}{e^{(1+\beta)\gamma} - 1}. \tag{3.12}$$

Based on (3.3) and (3.8):

$$\psi = \frac{\varepsilon}{\varepsilon + 1} \tag{3.13}$$

And finally, if ψ_p indicates the value of ψ for fluids in parallel flow,

$$\boxed{\psi_p = \frac{e^{-(1+\beta)\gamma} + \beta}{1 + \beta}} \tag{3.14}$$

If the fluids are in counter flow, instead of (3.1) we have:

$$q = US\frac{(t_1' - t_2'') - (t_2' - t_1'')}{\log_e \frac{t_1' - t_2''}{t_2' - t_1''}}. \tag{3.15}$$

Note that

$$\frac{t_1' - t_2''}{t_2' - t_1''} = \frac{\frac{t_1' - t_1''}{t_1' - t_2'} - \frac{t_2'' - t_1''}{t_1' - t_2'}}{\frac{t_1' - t_1''}{t_1' - t_2'} - 1}. \tag{3.16}$$

Assuming that

$$\eta = \frac{t_1' - t_1''}{t_1' - t_2'}, \tag{3.17}$$

and recalling (3.7) we have:

$$\frac{t_1' - t_2''}{t_2' - t_1''} = \frac{\eta - \beta}{\eta - 1}. \tag{3.18}$$

By analogy with (3.10) we also have:

$$\gamma = \frac{t_1' - t_2'}{(t_1' - t_2'') - (t_2' - t_1'')} \log_e \frac{t_1' - t_2''}{t_2' - t_1''} \tag{3.19}$$

And from that, recalling (3.4) as well as (3.7):

$$\gamma = \frac{1}{1 - \beta} \log_e \frac{\eta - \beta}{\eta - 1} \tag{3.20}$$

(3.20) leads to the following:

$$\eta = \frac{\beta - e^{(1-\beta)\gamma}}{1 - e^{(1-\beta)\gamma}} \tag{3.21}$$

Observing that

$$\psi = 1 - \frac{1}{\eta} \tag{3.22}$$

if ψ_c indicates the value of ψ for fluids in counter flow, we have

$$\boxed{\psi_c = \frac{1-\beta}{e^{(1-\beta)\gamma} - \beta}} \tag{3.23}$$

Note that the value of ψ_c is undetermined if $\beta = 1$; in that case, though, we have:

$$\Delta t_{ml} = t_2' - t_1'' = t_1' - t_2'' \tag{3.24}$$

Therefore, based on (3.6) and (3.15)

$$\eta_e M' c'_{pm} \left(t_1' - t_2'\right) = US \left(t_2' - t_1''\right); \tag{3.25}$$

then

$$\frac{t_1' - t_1''}{t_2' - t_1''} - 1 = \gamma. \tag{3.26}$$

Thus, recalling (3.3):

$$\frac{1}{\psi} - 1 = \gamma. \tag{3.27}$$

Finally,

$$\boxed{\psi_c = \frac{1}{\gamma + 1}} \tag{3.28}$$

Figure 3.1 was built based on (3.14) and Fig. 3.2 was built based on (3.23) and (3.28).

The values of ψ for fluids in parallel flow and counter flow are shown in Tables B.1 and B.2 in Appendix B.

Once the value of ψ is known, recalling (3.3) temperature t_2' is given by

$$t_2' = t_1'' + \psi \left(t_1' - t_1''\right). \tag{3.29}$$

If t_2' is known, based on (3.7) we have

$$t_2'' = t_1'' + \beta \left(t_1' - t_2'\right) \tag{3.30}$$

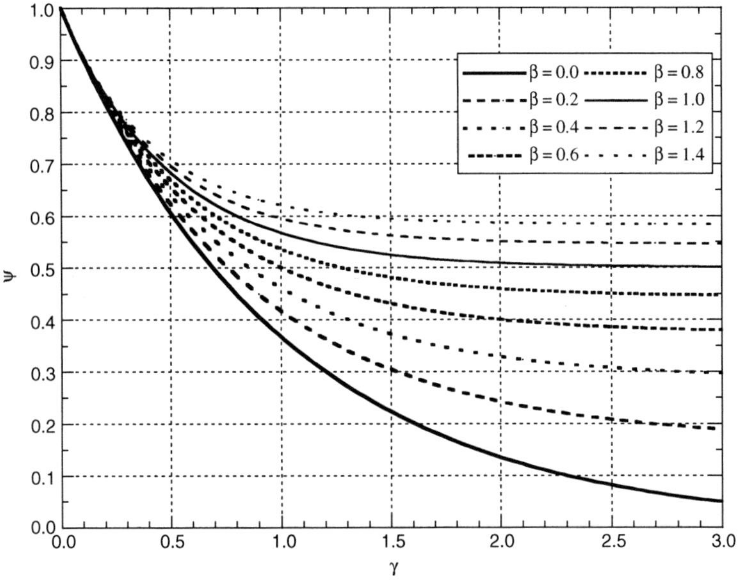

Fig. 3.1 Factor ψ for parallel flow

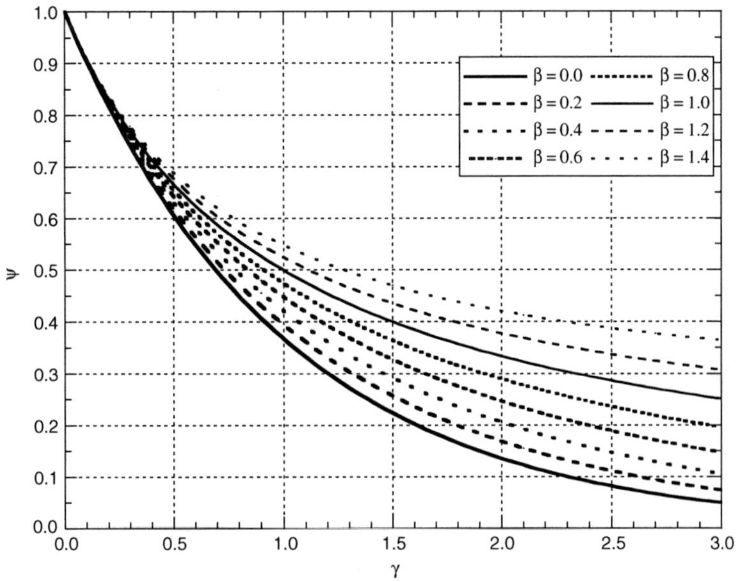

Fig. 3.2 Factor ψ for counterflow

As far as the exchange efficiency η_e, there are no values carrying general validity because it is influenced by numerous factors.

With heat exchangers its value depends on the heat loss on the outside. If the exchanger is well insulated it is possible to have very high values of η_e that are equal to 0.98–0.99; the same criterion is true for tube banks if they are contained in a casing.

The tube bank can be located in a space where its walls consist of tubes filled with another fluid; in that case the flue gas transfers part of the heat to these tubes, and the exchange efficiency referred to the bank can therefore be considerably lower than unity. In that case, though, there is no heat loss.

Sometimes, if the inlet temperature t'_1 of the heating fluid is known, the outlet temperature t''_2 of the heated fluid is imposed, while the inlet temperature t''_1 is unknown; in that case the latter temperature is calculated through the following equation

$$t''_1 = \frac{t''_2 - \beta t'_1 (1 - \psi)}{1 - \beta + \beta\psi}, \tag{3.31}$$

where temperature t'_2 is always calculated through (3.29).

It is certainly interesting to compare fluids in parallel flow and counter flow in relation to heat transfer.

Based on (3.5) and (3.6) we have:

$$q = \frac{US}{\gamma} \left(t'_1 - t'_2\right). \tag{3.32}$$

Based on (3.29) and after a series of steps we have:

$$q = \frac{US}{\gamma} \left(t'_1 - t''_1\right) (1 - \psi). \tag{3.33}$$

Based on (3.33) we establish that heat q is proportional to $(1 - \psi)$; thus, if two exchangers or tube banks are compared with one another with the same values of U, $S.\gamma, t'_1, t''_1$, one with the fluids in parallel flow and the other with the fluids in counter flow, the ratio between the transferred heats is given by (Table 3.1):

$$\frac{q_{\text{counter}}}{q_{\text{parallel}}} = \frac{(1 + \beta)\left(e^{(1-\beta)\gamma} - 1\right)}{\left(e^{(1-\beta)\gamma} - \beta\right)\left(1 - e^{-(1+\beta)\gamma}\right)}. \tag{3.34}$$

The choice among fluids in parallel flow or counter flow is irrelevant until the difference in transferred heat in both instances amounts to a few percentage points, and the ratio is consequently roughly below 1.02–1.03.

We establish that these values of the ratio are matched by decreasing values of γ with increases of β.

For tube banks of a steam generator this means that in that respect the values of γ decrease passing from an economizer to a superheater, and then to an air heater.

Table 3.1 – Ratio between transferred heat with fluids in counter flow and fluids in parallel flow

	γ				
β	0.2	0.4	0.6	0.8	1.0
0.2	1.0024	1.0085	1.0172	1.0276	1.0390
0.4	1.0047	1.0165	1.0331	1.0527	1.0742
0.6	1.0069	1.0240	1.0477	1.0755	1.1056
0.8	1.0090	1.0311	1.0611	1.0958	1.1330
1.0	1.0111	1.0377	1.0733	1.1137	1.1565
1.2	1.0131	1.0439	1.0842	1.1294	1.1763

But for these devices the value of U decreases, so that we can basically conclude based on (3.34) that the ratio S/M' is crucial for deciding the type of flow.

In other words, if the surface is modest with respect to the mass flow rate of the heating fluid, it is possible to opt for the solution in parallel flow without sensible drawback with respect to the heat transfer.

In conclusion, if the heated fluid is a boiling liquid, the assumption must be $c''_{pm} = \infty$. Based on (3.14) and (3.23) (in this case they coincide since we cannot speak of fluids in parallel or counter flow) we obtain

$$\psi = e^{-\gamma} \tag{3.35}$$

The value of U, included in γ, can be calculated with considerable satisfaction by referring for α' to the mean logarithmic temperature; this corresponds to what was pointed out earlier.

With heat exchangers it is customary to consider their efficiency. It is given by the ratio between the actual heat exchange and the maximum value of the heat that the exchanger could theoretically exchange. The latter corresponds to the infinite surface, so that we have $\gamma = \infty$.

Note that heat q transferred into the exchanger is equal to

$$q = \eta_e M' c'_{pm} \left(t'_1 - t''_1\right)(1 - \psi) \tag{3.36}$$

as can easily be verified.

In the case of fluids in parallel flow, with $\gamma = \infty$, based on (3.14) we obtain

$$1 - \psi_p = \frac{1}{1 + \beta}; \tag{3.37}$$

this is the maximum transferred heat indicated by q_∞ and equal to

$$q_\infty = \eta_e M' c'_{pm} \left(t'_1 - t''_1\right) \frac{1}{1 + \beta}. \tag{3.38}$$

Under these conditions, of course we have $t'_2 = t''_2$

Efficiency E of the exchanger is therefore equal to

$$E = \frac{q}{q_\infty} = (1+\beta)\left(1-\psi_p\right). \tag{3.39}$$

If the fluids are in counter flow it is important to distinguish the instances where $\beta \leq 1$ from those where $\beta > 1$.

If $\beta < 1$ based on (3.23), with $\gamma = \infty$, we obtain $\psi = 0$; similarly, if $\beta = 1$ based on (3.28) we obtain $\psi = 0$. Then

$$E = \frac{q}{q_\infty} = 1 - \psi_c. \tag{3.40}$$

If $\psi = 0$ based on (3.29) we determine that $t_2' = t_1''$; if $\beta = 1$ the temperatures of the two fluids are equal in every position.

If instead $\beta > 1$ based on (3.23), with $\gamma = \infty$, we obtain $(1-\psi) = 1/\beta$; therefore,

$$E = \frac{q}{q_\infty} = \beta\left(1-\psi_c\right). \tag{3.41}$$

With $\gamma = \infty$ based on (3.29) and (3.30) through a series of steps we obtain

$$t_2'' = t_1'$$

Table 3.2 Heat exchanger efficiency

Fluids in parallel flow

β	Y 0.5	1.0	1.5	2.0	2.5	3.0
0.5	0.528	0.777	0.895	0.950	0.976	0.989
1.0	0.632	0.865	0.950	0.982	0.993	0.998
1.5	0.713	0.918	0.976	0.993	0.998	0.999
2.0	0.777	0.950	0.989	0.998	0.999	0.999

Fluids in counter flow

β	Y 0.5	1.0	1.5	2.0	2.5	3.0
0.5	0.362	0.565	0.691	0.775	0.833	0.874
1.0	0.333	0.500	0.600	0.667	0.714	0.750
1.5	0.460	0.661	0.770	0.838	0.882	0.913
2.0	0.565	0.775	0.874	0.927	0.957	0.974

Since ψ_p and ψ_c are a function of β and γ, it is possible to build Table 3.2 showing the values of E for various values of these two parameters for fluids in parallel flow, as well as counter flow.

3.3 Factor ψ in Real Cases

For fluids in cross flow and for all the instances involving heat exchangers we will examine, we will compute the actual values of ψ by including them in the various tables.

But for coils and tube banks, given that their behavior resembles that of fluids in parallel flow or counter flow, we will refer to these two situations by introducing corrective factors to obtain the actual value of ψ.

By recalling that ψ_p and ψ_c are the values of ψ in reference to fluids in parallel flow or counter flow, if we refer to fluids in parallel flow the actual value of ψ is given by

$$\boxed{\psi = \varphi_p \psi_p} \tag{3.42}$$

The values of the corrective factor φ_p are listed in some tables of Appendix B.

In reference to fluids in counter flow instead we will set

$$\boxed{\psi = \varphi_c \psi_c} \tag{3.43}$$

and listing in some tables of Appendix B corrective factor φ_c.

ψ_p is obtained through (3.14), whereas ψ_c is calculated through (3.23) and (3.28). Factors β and γ included in these equations are computed through (3.4) and (3.5). As we shall see, β and γ are crucial to obtain the corrective factors, as well.

3.3.1 Fluids with Cross Flow

For fluids with cross flow (Fig. 2.4) we directly calculated factor ψ; its values are shown in Table B.3 in Appendix B.

If $\beta \leq 2.3$ and $\gamma \leq 2$ factor ψ for cross flow can also be computed with excellent approximation through the following equation:

$$\psi = (1 - Z)\psi_c + Z\psi_p \tag{3.42}$$

with

$$Z = 0.5 - 0.136\,(1 + 0.24\beta)\sqrt[3]{\gamma} \tag{3.43}$$

As you see, the value of ψ is intermediate between ψ_p and ψ_c. The behavior of fluids with cross flow is therefore intermediate between the two classic ways, and slightly closer to the condition of fluids in counter flow.

3.3.2 Heat Exchangers

3.3.2.1 Heat Exchangers with Two Passages of the Internal Fluid

This is in reference to heat exchangers with two passages of the fluid inside the tubes schematized in Fig. 2.5.

As stated earlier, the tables directly show the values of ψ relative to various types of exchangers; this facilitates the verification computation.

The references to various types of heat exchangers are identical to the ones adopted for the design computation in Sect. 2.3.2.1.

This is the examined range: $\beta = 0.1$–3.0 and $\gamma = 0.05$–4.0.

The values of ψ for types A and D or for types B and C with two passages of the external fluid are shown in Tables B.4 and B.5.

The values of ψ for types A and B or for types C and D with three passages of the external fluid are shown in Tables B.6 and B.7.

The values of ψ for types A and D or for types B and C with four passages of the external fluid are shown in Tables B.8 and B.9.

Finally, the values of ψ for types A and B or for types C and D with five passages of the external fluid are shown in Tables B.10 and B.11.

Note that in Tables B.4, B.6, B.8 and B.10 within the examined range of γ, factor ψ shoes a minimum for almost all values of β; therefore, it is impossible to obtain a lower value of ψ for the considered value of β.

Note that the behavior of these exchangers is close to that of fluids in parallel flow.

3.3.2.2 Heat Exchangers with Three Passages of the Internal Fluid

This is in reference to heat exchangers with three passages of the fluid inside the tubes schematized in Fig. 2.6.

Even in this case the tables directly show the values of factor ψ.

The references to the various types of heat exchangers are identical to those adopted for the design computation in Sect. 2.3.2.2.

This is the examined range: $\beta = 0.1 - 3.0$ and $\gamma = 0.05 - 4.0$.

The values of ψ for types E and F or for types G and H with two passages of the external fluid are shown in Tables B.12 and B.13.

The values of ψ for type E or for type G with three passages of the external fluid are shown in Tables B.14 and B.15.

The values of ψ for types E and F or for types G and H with four passages of the external fluid are shown in Tables B.16 and B.17.

Finally, the values of ψ for type E or for type G with five passages of the external fluid are shown in Tables B.18 and B.19.

Please, note that in Tables B.12, B.14, B.16 and B.18 within the examined range of γ factor γ shows a minimum for almost all values of β.

Note that the behavior of these exchangers is close to that of fluids in parallel flow.

3.3.2.3 Heat Exchangers with Four Passages of the Internal Fluid

This is in reference to heat exchangers with four passages of the fluid inside the tubes schematized in Fig. 2.7.

Even in this case the Tables show directly the values of factor ψ.

The reference to various types of heat exchangers are identical to those adopted for the design computation in Sect. 2.3.2.3.

This is the examined range: $\beta = 0.1$–3.0 and $\gamma = 0.05$–4.0.

The values of ψ for types I and N or for types L and M with two passages of the external fluid are shown in Tables B.20 and B.21.

The values of ψ for types M and N with three passages of the external fluid are shown in Table B.22.

The values of ψ for types L and M with four passages of the external fluid are shown in Table B.23.

Finally, the values of ψ for types M and N with five passages of the external fluid are shown in Table B.24.

The values of ψ for the following types of exchangers are not shown : types I and L with three passages of the external fluid, types I and N with four passages of the external fluid, types I and L with five passages of the external fluid.

As pointed out earlier, for design computation the behavior of these exchangers is quite close to that of similar exchangers with two instead of four passages of the internal fluid.

As far as the values of ψ for these exchangers it is possible to refer to Tables B.6, B.8 and B.10.

As already pointed out for exchangers with two or three passages of the internal fluid, the value of ψ shows a minimum for almost all values of β even in Table B.20; even in this case note that the exchangers in question behave in a similar way to fluids in parallel flow.

3.3.3 Coils

Coils shown in Fig. 2.8 are usually considered, as already pointed out in Sect. 2.3.3, coils with fluids in parallel flow. By analogy, those in Fig. 2.9 are considered coils with fluids in counter flow.

Based on this consideration and contrary to the case with exchangers, we preferred to refer to the value of ψ relative to fluids in paralled flow or in counter flow by listing the necessary corrective factors φ in the tables.

Recalling that heat transfer is proportional to $(1 - \psi)$, we ignored the cases where the difference between the actual heat transfer and the one relative to fluids in parallel flow or counter flow (depending on the type of coil) is below $\pm 1\%$, given that this difference can be considered insignificant.

We consider the following range: $\beta = 0.1$–3.0 e $\gamma = 0.05$–4.0.

3.3.3.1 Coils with Fluids in Parallel Flow

Since the reference is to fluids in parallel flow the corrective factor shown in the tables is φ_p.

Let us consider the coil in Fig. 2.8; if it consists of 2 sections the corrective factor φ_p can be obtained through Table B.25.

As you see, in some cases (the most likely ones) the value of the corrective factor is below unity which means that the heat transfer is greater than that with pure parallel flow.

In other instances the opposite is true; this happens for high values of β and γ; therefore, for these the heat transfer is less favorable in comparison with fluids in parallel flow.

Table B.26 shows the values of φ_p relative to a coil with 3 sections; the values are always below unity and its behavior is always better than fluids in parallel flow.

If we consider coils with 3 or more sections, we see that the difference in transferred heat in the coil with respect to fluids in parallel flow is slightly greater than $\pm$ 1% for few cases with high values (unlikely) of β and γ. Considering that a coil with only two sections is exceptional and unlikely, it is possible to state that coils in parallel flow behave in fact as such with regard to the value of ψ in verification calculation.

3.3.3.2 Coils with Fluids in Counter Flow

Let us now examine coils in counter flow in Fig. 2.9.

Since the reference is to fluids in counter flow the corrective factor shown in the Tables is φ_c.

The values of φ_c in Tables B.27, B.28, B.29 and B.30 with reference to coils with 2, 3, 4 and 6 sections.

As expected, we establish that the corrective factor is always greater than unity, and this means that the transferred heat is less than that corresponding to fluids in actual counter flow.

We see that there are less and less cases to consider when going from 2 to 6 sections. For a number of sections equal to 6 or more the difference in heat transferred in the coil and in fluids in counter flow slightly exceeds 1% only in few and rather exceptional instances with high values of γ.

In conclusion, for a number of sections equal or higher than 6 (as is usually the case) the verification calculation conducted for fluids in counter flow is correct, whereas for a smaller number of sections, if necessary, one can introduce the indicated corrective factor.

3.3.4 Tube Bank with Various Passages of the External Fluid

We refer to the devices in Figs. 2.10 and 2.11. As already pointed out in Sect. 2.3.4, the classic device of this kind is the heat recuperator located at the end of a steam generator, provided that this type of exchanger can be used also with other fluids, typically gaseous fluids.

The external fluid can enter the tube bank in correspondence of the entrance into the tubes of the internal fluid, or the opposite takes place with the external fluid entering the tube bank in correspondence of the exit from the tubes of the internal fluid. Figure 2.10 represents a tube bank of the first kind with three passages of the external fluid. Figure 2.11 represents a tube bank of the second kind instead.

Again, in the first case we speak of a tube bank with fluids in parallel flow, and in the second case we speak of a tube bank with fluids in counter flow.

As for coils, we will refer to ψ_p for the first type, and to ψ_c for the second one by introducing the usual corrective factors φ_p and φ_c.

This is the considered range: $\beta = 0.1-3.0$ e $\gamma = 0.05-4.0$

3.3.4.1 Tube Banks with Fluids in Parallel Flow

As far as the tube bank with fluids in parallel flow in Fig. 2.10, Tables B.31 and B.32 list the values of φ_p relative to 2 and 3 passages of the external fluid.

In Table B.31 the values of φ_p are below unity for low values of β and γ. This means that the heat transfer is greater than that corresponding to fluids in parallel flow. Instead, for high values of β and γ the corrective factor is greater than unity. For these values the behavior of the tube bank is worse compared to that of fluids in parallel flow.

If the number of passages of the external fluid is equal to 3 instead (Table B.32) the corrective factor φ_p is always below unity, applies to a limited number of cases and is generally close to unity.

Thus, it is possible to conclude that if the number of passages of the external fluid is ≥ 3, the tube bank may be considered as actually having fluids in parallel flow as far as the verification computation, thus neglecting the potential small advantage represented by the fact that $\varphi_p < 1$.

3.3.4.2 Tube Banks with Fluids in Counter Flow

Given the reference to fluids in counter flow the corrective factor in the tables is φ_c.

Tables B.33, B.34 and B.35 are about tube banks with fluids in counter flow shown in Fig. 2.11, respectively with 2, 3 and 4 passages of the external fluid.

We did not consider a higher number of passages because it is unlikely.

Note that the values of φ_c are always above unity which means that the transferred heat into tube bank is less than the one occurring with the fluids in actual counter flow.

In addition, in the case of 4 passages of the external fluid (Table B.35) the corrective factor must be adopted for a reduced number of cases but it may be considerably higher than unity.

In conclusion, contrary to what was observed for tube banks with fluids in parallel flow, even with 4 passages of the external fluid, it is best to ensure that the behavior of the tube bank is not very dissimilar to that of fluids in counter flow.

Appendix A
Corrective Factors for Design Computation

A.1 Fluids in Cross Flow

Fig. A.1

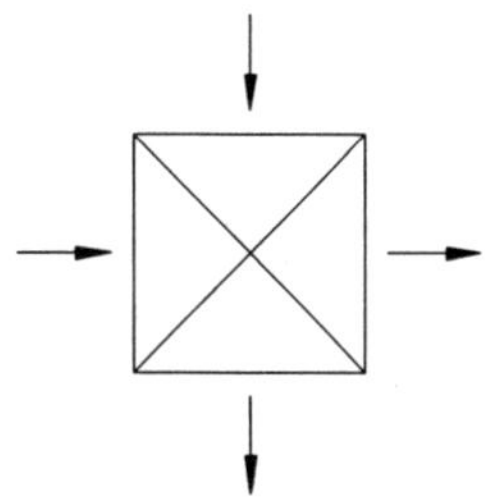

Table A.1 – Fluids in cross flow – Corrective factor χ_c

β \ ψ	0.03	0.06	0.09	0.12	0.15	0.18	0.21	0.24	0.27	0.30	0.33
0.1	**0.899**	*0.925*	*0.940*	*0.950*	*0.958*	0.965	0.969	0.974	0.977	0.981	0.983
0.2	**0.802**	*0.848*	*0.876*	*0.897*	*0.914*	*0.926*	0.937	0.946	0.953	0.959	0.965
0.3	....	**0.771**	*0.811*	*0.841*	*0.865*	*0.884*	*0.901*	0.914	0.926	0.936	0.945
0.4	....	**0.695**	**0.743**	*0.782*	*0.813*	*0.839*	*0.861*	*0.880*	0.896	0.910	0.923
0.5	......	....	**0.674**	**0.720**	*0.758*	*0.790*	*0.818*	*0.842*	0.864	0.882	0.899
0.6	......	....	....	**0.655**	**0.699**	*0.737*	*0.771*	*0.801*	*0.827*	0.851	0.871
0.7	......	....	....	.....	**0.635**	**0.680**	*0.720*	*0.756*	*0.788*	*0.816*	0.841
0.8	......	....	....	....	....	**0.618**	**0.664**	*0.706*	*0.743*	*0.777*	*0.808*
0.9	......	....	....	....	....	....	**0.601**	**0.650**	*0.694*	*0.734*	*0.770*
1.0	......	....	....	....	....	....	....	**0.587**	**0.638**	*0.685*	*0.728*
1.1	....	....	....	....	....	....	....	....	**0.573**	**0.629**	*0.679*
1.2	....	....	....	....	....	....	....	....	....	**0.561**	**0.622**
1.3	....	....	....	....	....	....	....	....	....	....	**0.552**

If β >1.3 no solution is possible in this field
Values in italics : 2<γ<4; **Values in bold type : 4< γ <6**

D. Annaratone, *Handbook for Heat Exchangers and Tube Banks design*,
DOI 10.1007/978-3-642-13309-1, © Springer-Verlag Berlin Heidelberg 2010

Table A.1 – (continued)

	ψ										
β	0.33	0.36	0.39	0.42	0.45	0.48	0.51	0.54	0.57	0.60	0.63
0.1	0.984	0.986	0.988	0.989	....	....	....	....	....	....	....
0.2	0.965	0.969	0.974	0.977	0.981	0.984	0.986	0.988	....	....	....
0.3	0.945	0.953	0.959	0.965	0.970	0.974	0.978	0.982	0.985	0.987	0.990
0.4	0.923	0.933	0.943	0.951	0.958	0.964	0.970	0.974	0.979	0.982	0.986
0.5	0.899	0.913	0.925	0.936	0.945	0.954	0.961	0.967	0.972	0.977	0.981
0.6	0.871	0.889	0.905	0.919	0.931	0.942	0.951	0.959	0.966	0.972	0.977
0.7	0.841	0.864	0.883	0.900	0.915	0.929	0.940	0.950	0.958	0.966	0.972
0.8	*0.808*	0.835	0.859	0.880	0.898	0.914	0.928	0.940	0.951	0.959	0.967
0.9	*0.770*	*0.803*	0.832	0.857	0.879	0.899	0.915	0.930	0.942	0.953	0.962
1.0	*0.728*	*0.766*	0.801	0.831	0.857	0.881	0.901	0.918	0.933	0.946	0.956
1.1	*0.679*	*0.725*	*0.765*	0.802	0.833	0.861	0.885	0.905	0.923	0.937	0.950
1.2	**0.622**	*0.676*	*0.725*	*0.768*	0.806	0.839	0.867	0.891	0.911	0.929	0.943
1.3	**0.552**	**0.619**	*0.678*	*0.729*	0.774	0.813	0.846	0.875	0.899	0.919	0.936
1.4	....	**0.545**	*0.620*	*0.683*	*0.737*	0.784	0.823	0.857	0.885	0.908	0.927
1.5	....	....	**0.544**	*0.626*	*0.693*	*0.749*	0.797	0.836	0.869	0.896	0.919
1.6	....	....	....	**0.548**	*0.638*	*0.708*	0.765	0.813	0.851	0.883	0.909
1.7	....	....	....	....	**0.561**	*0.656*	*0.728*	0.785	0.831	0.868	0.898
1.8	....	....	....	....	....	**0.584**	*0.681*	0.752	0.808	0.851	0.886
1.9	....	....	....	....	....	....	*0.617*	*0.711*	0.780	0.832	0.872
2.0	....	....	....	....	....	....	....	*0.657*	0.746	0.809	0.857
2.1	....	....	....	....	....	....	....	**0.573**	*0.702*	0.782	0.839
2.2	....	....	....	....	....	....	....	....	*0.641*	0.748	0.818
2.3	....	....	....	....	....	....	....	....	**0.519**	*0.704*	0.793
2.4	....	....	....	....	....	....	....	....	....	*0.638*	0.762
2.5	....	....	....	....	....	....	....	....	....	....	0.721
2.6	....	....	....	....	....	....	....	....	....	....	*0.659*

If $\beta > 2.6$ no solution is possible in this field

Values in italics : $2<\gamma<4$; **Values in bold type : $4<\gamma<6$**

Table A.1 – (continued)

β	ψ 0.63	0.66	0.69	0.72	0.75	0.78	0.81	0.84	0.87
0.3	0.990	....	....	....	....	....	....	....	....
0.4	0.986	0.988	....	....	....	....	....	....	....
0.5	0.981	0.985	0.988	....	....	....	....	....	....
0.6	0.977	0.981	0.985	0.989	....	....	....	....	....
0.7	0.972	0.978	0.982	0.986	0.990	....	....	....	....
0.8	0.967	0.974	0.979	0.984	0.988	....	....	....	....
0.9	0.962	0.969	0.976	0.981	0.986	0.990	....	....	....
1.0	0.956	0.965	0.972	0.979	0.984	0.988	....	....	....
1.1	0.950	0.960	0.969	0.976	0.982	0.987	....	....	....
1.2	0.943	0.955	0.965	0.973	0.980	0.985	0.990	....	....
1.3	0.936	0.949	0.961	0.970	0.978	0.984	0.989	....	....
1.4	0.927	0.943	0.956	0.967	0.975	0.982	0.988	....	....
1.5	0.919	0.937	0.952	0.963	0.973	0.980	0.986	....	....
1.6	0.909	0.930	0.947	0.960	0.970	0.979	0.985	....	....
1.7	0.898	0.922	0.941	0.956	0.968	0.977	0.984	0.990	....
1.8	0.886	0.914	0.935	0.952	0.965	0.975	0.983	0.989	....
1.9	0.872	0.904	0.928	0.947	0.962	0.973	0.981	0.988	....
2.0	0.857	0.893	0.921	0.942	0.958	0.971	0.980	0.987	....
2.1	0.839	0.881	0.913	0.937	0.955	0.969	0.979	0.986	....
2.2	0.818	0.868	0.904	0.931	0.951	0.966	0.977	0.985	....
2.3	0.793	0.852	0.895	0.925	0.947	0.964	0.975	0.984	....
2.4	0.762	0.834	0.883	0.918	0.943	0.961	0.974	0.983	....
2.5	0.721	0.813	0.871	0.910	0.938	0.958	0.972	0.982	0.990
2.6	*0.659*	0.787	0.856	0.902	0.933	0.955	0.970	0.981	0.989
2.7	....	0.752	0.840	0.893	0.928	0.952	0.969	0.980	0.988
2.8	....	*0.704*	0.819	0.882	0.922	0.948	0.967	0.979	0.988
2.9	....	*0.616*	0.794	0.870	0.915	0.945	0.965	0.978	0.987
3.0	....	....	0.762	0.856	0.908	0.941	0.962	0.977	0.987

If β <0.3 or ψ >0.87 use the mean logarithmic temperature difference for counter flow
Values in italics : 2<γ<4

A.2 Heat Exchangers

A.2.1 Heat Exchangers with 2 Passages of Internal Fluid (Fig. 2.5)

Fig. A.2

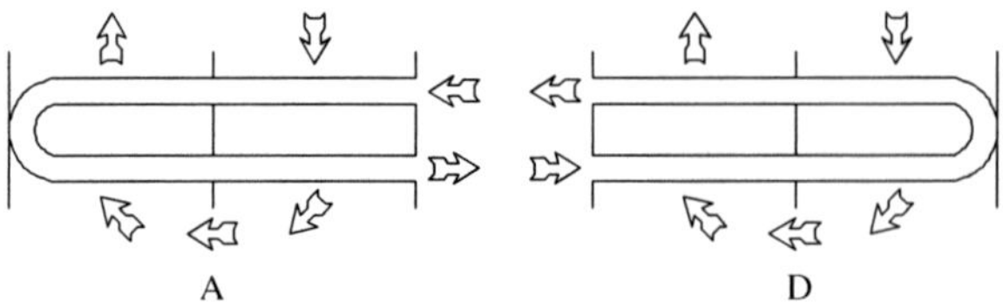

Table A.2 – Heat exchangers – A and D type – 2 passages of external fluid Corrective factor χ_c

β \ ψ	0.08	0.12	0.16	0.20	0.24	0.28	0.32	0.36	0.40	0.44
0.05	*0.885*	*0.933*	0.955	0.967	0.975	0.981	0.985	0.988	....	....
0.10	**0.638**	*0.841*	*0.899*	0.928	0.947	0.960	0.969	0.976	0.981	0.985
0.15	....	*0.681*	*0.824*	0.881	0.914	0.936	0.951	0.962	0.970	0.976
0.20	....	....	*0.711*	*0.821*	0.875	0.908	0.930	0.946	0.958	0.967
0.25	....	....	....	*0.737*	0.826	0.875	0.906	0.929	0.945	0.958
0.30	....	....	....	*0.565*	*0.762*	0.835	0.879	0.909	0.931	0.947
0.35	....	....	....	....	*0.666*	0.786	0.847	0.887	0.915	0.935
0.40	....	....	....	....	....	*0.719*	0.809	0.861	0.897	0.922
0.45	....	....	....	....	....	*0.609*	0.760	0.831	0.876	0.907
0.50	....	....	....	....	....	....	*0.694*	0.795	0.852	0.891
0.60	....	....	....	....	....	....	....	0.687	0.792	0.852
0.70	....	....	....	....	....	....	....	....	0.698	0.799
0.80	....	....	....	....	....	....	....	....	....	0.721
0.90	....	....	....	....	....	....	....	....	....	*0.551*

If β >0.9 or ψ <0.08 no solution is possible in this field
Values in italics : 2<γ< 4 ; **Value in bold type : 4<γ<6**

Table A.2 – (continued)

	ψ										
β	0.48	0.52	0.56	0.60	0.64	0.68	0.72	0.76	0.80	0.84	0.88
0.1	0.988	....	....	....	....	....	....	....	....	....	....
0.2	0.975	0.980	0.985	0.988	....	....	....	....	....	....	....
0.3	0.959	0.968	0.976	0.982	0.986	0.990	....	....	....	....	....
0.4	0.940	0.955	0.966	0.974	0.981	0.986	....	....	....	....	....
0.5	0.919	0.939	0.954	0.966	0.975	0.982	0.987	....	....	....	....
0.6	0.892	0.920	0.941	0.956	0.968	0.977	0.984	0.989	....	....	....
0.7	0.858	0.897	0.926	0.946	0.961	0.972	0.981	0.987	....	....	....
0.8	0.814	0.870	0.907	0.934	0.953	0.967	0.977	0.985	....	....	....
0.9	0.753	0.835	0.885	0.919	0.943	0.961	0.973	0.982	0.989	....	....
1.0	0.651	0.789	0.859	0.903	0.933	0.954	0.969	0.980	0.987	....	....
1.1	....	0.721	0.825	0.883	0.921	0.947	0.964	0.977	0.986	....	....
1.2	....	0.591	0.779	0.859	0.907	0.938	0.959	0.974	0.984	....	....
1.3	....	....	0.713	0.829	0.890	0.929	0.954	0.971	0.982	0.990	....
1.4	....	....	0.581	0.790	0.871	0.918	0.948	0.967	0.980	0.989	....
1.5	....	....	....	0.734	0.847	0.905	0.941	0.963	0.978	0.988	....
1.6	....	....	....	0.640	0.816	0.891	0.933	0.959	0.976	0.987	....
1.7	....	....	....	....	0.776	0.874	0.925	0.955	0.974	0.986	....
1.8	....	....	....	....	0.719	0.853	0.915	0.950	0.971	0.985	....
1.9	....	....	....	....	0.613	0.827	0.904	0.945	0.969	0.983	....
2.0	....	....	....	....	....	0.794	0.891	0.939	0.966	0.982	....
2.1	....	....	....	....	....	0.748	0.875	0.932	0.963	0.980	....
2.2	....	....	....	....	....	0.677	0.857	0.925	0.960	0.979	....
2.3	....	....	....	....	....	....	0.835	0.916	0.956	0.977	0.989
2.4	....	....	....	....	....	....	0.807	0.907	0.952	0.976	0.989
2.5	....	....	....	....	....	....	0.769	0.896	0.948	0.974	0.988
2.6	....	....	....	....	....	....	0.715	0.884	0.943	0.972	0.987
2.7	....	....	....	....	....	....	0.613	0.869	0.938	0.970	0.987
2.8	....	....	....	....	....	....	....	0.851	0.933	0.969	0.986
2.9	....	....	....	....	....	....	....	0.830	0.927	0.966	0.985
3.0	....	....	....	....	....	....	....	0.802	0.920	0.964	0.985

If ψ >0.88 use the mean logarithmic temperature difference for counter flow

Fig. A.3

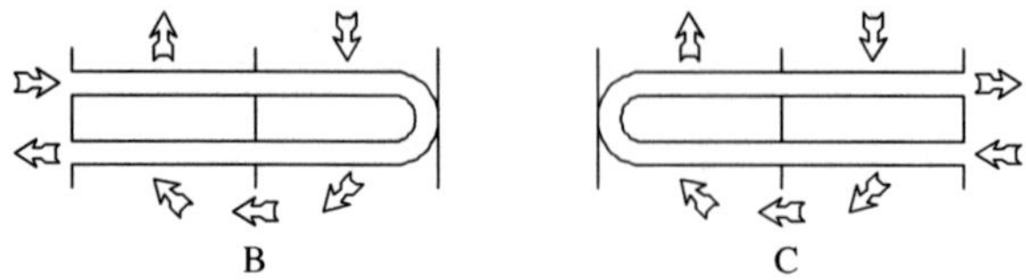

Table A.3 – Heat exchangers – B and C type – 2 passages of external fluid Corrective factor χ_c

β \ ψ	0.04	0.08	0.12	0.16	0.20	0.24	0.28	0.32	0.36	0.40	0.44
0.10	**0.759**	*0.875*	*0.919*	*0.943*	0.957	0.967	0.975	0.980	0.984	0.987	....
0.15	**0.607**	*0.793*	*0.868*	*0.908*	0.932	0.948	0.960	0.969	0.975	0.980	0.985
0.20	....	**0.696**	*0.808*	*0.867*	0.903	0.927	0.943	0.956	0.965	0.973	0.978
0.30	....	....	*0.656*	*0.764*	*0.831*	0.875	0.905	0.927	0.943	0.956	0.965
0.40	....	....	**0.484**	*0.628*	*0.735*	*0.807*	0.856	0.890	0.916	0.935	0.950
0.50	....	....	....	**0.476**	*0.609*	*0.716*	*0.792*	0.845	0.882	0.910	0.931
0.60	....	....	....	....	**0.470**	*0.599*	*0.707*	0.785	0.840	0.880	0.910
0.70	....	....	....	....	....	**0.470**	*0.598*	*0.707*	0.786	0.842	0.883
0.80	....	....	....	....	....	....	**0.476**	*0.606*	*0.715*	0.794	0.850
0.90	....	....	....	....	....	....	....	*0.490*	*0.623*	0.732	0.808
1.00	....	....	....	....	....	....	....	**0.380**	*0.514*	*0.650*	0.755
1.10	....	....	....	....	....	....	....	....	**0.405**	*0.548*	0.685
1.20	....	....	....	....	....	....	....	....	....	**0.441**	*0.595*
1.30	....	....	....	....	....	....	....	....	....	**0.342**	*0.491*
1.40	....	....	....	....	....	....	....	....	....	....	**0.391**

If β >1.4 no solution is possible in this field

Values in italics : $2<\gamma<4$; **Values in bold type :** $4<\gamma<6$

Table A.3 – (continued)

	ψ										
β	0.48	0.52	0.56	0.60	0.64	0.68	0.72	0.76	0.80	0.84	0.88
0.2	0.983	0.987	0.990	....	....	....	....	....	....	....	....
0.3	0.973	0.979	0.984	0.988	....	....	....	....	....	....	....
0.4	0.961	0.970	0.977	0.983	0.987	....	....	....	....	....	....
0.5	0.948	0.960	0.970	0.977	0.983	0.988	....	....	....	....	....
0.6	0.932	0.948	0.961	0.971	0.979	0.985	0.989	....	....	....	....
0.7	0.913	0.935	0.951	0.964	0.974	0.981	0.987	....	....	....	....
0.8	0.890	0.919	0.940	0.957	0.969	0.978	0.985	0.990	....	....	....
0.9	0.862	0.900	0.928	0.948	0.963	0.974	0.982	0.988	....	....	....
1.0	0.827	0.878	0.913	0.938	0.956	0.969	0.979	0.986	....	....	....
1.1	0.783	0.850	0.895	0.926	0.949	0.965	0.976	0.985	....	....	....
1.2	0.726	0.815	0.874	0.913	0.940	0.959	0.973	0.982	0.989	....	....
1.3	*0.651*	0.771	0.848	0.897	0.931	0.954	0.969	0.980	0.988	....	....
1.4	*0.558*	0.713	0.815	0.879	0.920	0.947	0.965	0.978	0.987	....	....
1.5	*0.456*	0.638	0.774	0.856	0.907	0.940	0.961	0.976	0.985	....	....
1.6	**0.361**	*0.545*	0.720	0.828	0.892	0.931	0.956	0.973	0.984	....	....
1.7	....	*0.446*	0.650	0.793	0.874	0.922	0.951	0.970	0.982	....	....
1.8	....	**0.352**	*0.561*	0.748	0.852	0.911	0.945	0.967	0.981	0.989	....
1.9	....	....	*0.464*	0.689	0.826	0.898	0.939	0.964	0.979	0.989	....
2.0	....	....	**0.371**	0.611	0.792	0.883	0.931	0.960	0.977	0.988	....
2.1	....	....	....	*0.518*	0.749	0.864	0.923	0.956	0.975	0.987	....
2.2	....	....	....	*0.423*	0.692	0.842	0.914	0.952	0.973	0.986	....
2.3	....	....	....	**0.331**	0.616	0.815	0.903	0.947	0.971	0.985	....
2.4	....	....	....	....	*0.525*	0.781	0.890	0.941	0.969	0.984	....
2.5	....	....	....	....	*0.431*	0.735	0.875	0.935	0.966	0.983	....
2.6	....	....	....	....	**0.338**	0.675	0.856	0.928	0.963	0.982	....
2.7	....	....	....	....	....	0.597	0.834	0.921	0.960	0.980	....
2.8	....	....	....	....	....	*0.504*	0.806	0.912	0.957	0.979	....
2.9	....	....	....	....	....	*0.410*	0.770	0.902	0.953	0.978	....
3.0	....	....	....	....	....	**0.313**	0.723	0.890	0.949	0.976	0.989

If β <0.2 or ψ >0.88 use the mean logarithmic temperature difference for counter flow
Values in italics : 2<γ<4 ; **Values in bold type : 4<γ<6**

Fig. A.4

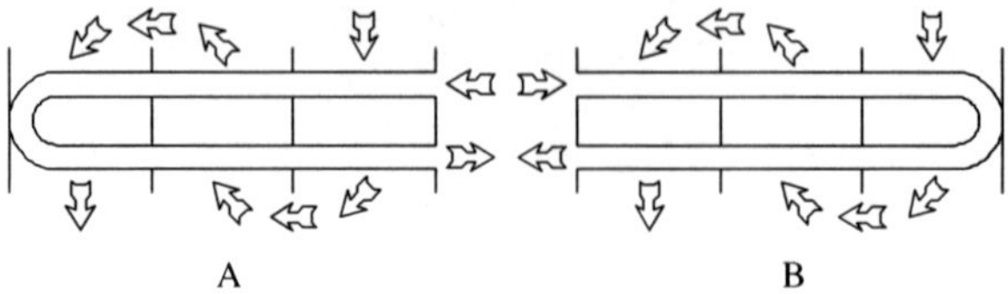

Table A.4 – Heat exchangers – A and B type – 3 passages of external fluid Corrective factor χ_c (valid also for I and L type with 3 passages of external fluid)

	ψ										
β	0.04	0.08	0.12	0.16	0.20	0.24	0.28	0.32	0.36	0.40	0.44
0.05	**0.723**	*0.901*	*0.941*	0.960	0.971	0.978	0.983	0.987	0.989	....	....
0.10	....	*0.736*	*0.864*	*0.911*	0.937	0.953	0.964	0.972	0.978	0.983	0.986
0.15	....	....	*0.750*	*0.849*	0.896	0.924	0.942	0.956	0.965	0.973	0.979
0.20	....	....	....	*0.764*	*0.846*	0.890	0.918	0.937	0.952	0.962	0.970
0.25	....	....	....	*0.616*	*0.781*	0.849	0.899	0.917	0.936	0.950	0.962
0.30	....	....	....	....	*0.686*	*0.798*	0.856	0.893	0.919	0.938	0.952
0.35	....	....	....	....	....	*0.730*	0.815	0.866	0.899	0.923	0.941
0.40	....	....	....	....	....	*0.624*	*0.765*	0.833	0.877	0.908	0.930
0.45	....	....	....	....	....	....	*0.696*	0.795	0.851	0.890	0.917
0.50	....	....	....	....	....	....	*0.581*	0.746	0.821	0.869	0.902
0.60	....	....	....	....	....	....	....	*0.567*	0.740	0.819	0.869
0.70	....	....	....	....	....	....	....	....	*0.585*	0.747	0.824
0.80	....	....	....	....	....	....	....	....	....	*0.622*	0.764
0.90	....	....	....	....	....	....	....	....	....	....	0.667

If $\beta > 0.9$ no solution is possible in this field

Values in italics : $2<\gamma<4$; **Value in bold type : $4<\gamma<6$**

Table A.4 – (continued)

β \ ψ	0.48	0.52	0.56	0.60	0.64	0.68	0.72	0.76	0.80	0.84	0.88
0.1	0.989	....	....	....	....	....	....	....	....	....	....
0.2	0.977	0.982	0.986	0.989	....	....	....	....	....	....	....
0.3	0.963	0.971	0.978	0.983	0.987	....	....	....	....	....	....
0.4	0.946	0.959	0.969	0.976	0.983	0.987	....	....	....	....	....
0.5	0.927	0.945	0.958	0.969	0.977	0.984	0.988	....	....	....	....
0.6	0.903	0.928	0.946	0.961	0.971	0.979	0.986	....	....	....	....
0.7	0.874	0.908	0.933	0.951	0.965	0.975	0.983	0.988	....	....	....
0.8	0.837	0.884	0.917	0.940	0.957	0.970	0.979	0.986	....	....	....
0.9	0.788	0.854	0.898	0.928	0.949	0.964	0.976	0.984	0.990	....	....
1.0	0.716	0.816	0.874	0.913	0.940	0.958	0.972	0.982	0.988	....	....
1.1	0.580	0.763	0.846	0.896	0.929	0.952	0.968	0.979	0.987	....	....
1.2	....	0.684	0.808	0.875	0.916	0.944	0.963	0.976	0.986	....	....
1.3	....	....	0.758	0.849	0.902	0.936	0.958	0.973	0.984	....	....
1.4	....	....	0.682	0.817	0.885	0.926	0.953	0.970	0.982	0.990	....
1.5	....	....	....	0.774	0.864	0.915	0.947	0.967	0.980	0.989	....
1.6	....	....	....	0.712	0.839	0.902	0.940	0.963	0.978	0.988	....
1.7	....	....	....	0.600	0.807	0.888	0.932	0.959	0.976	0.987	....
1.8	....	....	....	....	0.764	0.870	0.923	0.955	0.974	0.986	....
1.9	....	....	....	....	0.700	0.848	0.914	0.950	0.972	0.985	....
2.0	....	....	....	....	0.576	0.821	0.902	0.945	0.969	0.984	....
2.1	....	....	....	....	....	0.786	0.889	0.939	0.966	0.982	....
2.2	....	....	....	....	....	0.737	0.874	0.932	0.963	0.981	....
2.3	....	....	....	....	....	0.659	0.855	0.925	0.960	0.980	....
2.4	....	....	....	....	....	....	0.832	0.917	0.957	0.978	0.990
2.5	....	....	....	....	....	....	0.803	0.907	0.953	0.976	0.989
2.6	....	....	....	....	....	....	0.764	0.897	0.949	0.975	0.989
2.7	....	....	....	....	....	....	0.707	0.884	0.945	0.973	0.988
2.8	....	....	....	....	....	....	0.599	0.869	0.940	0.971	0.987
2.9	....	....	....	....	....	....	....	0.851	0.934	0.969	0.987
3.0	....	....	....	....	....	....	....	0.829	0.928	0.967	0.986

If ψ>0.88 use the mean logarithmic temperature difference for counter flow

Fig. A.5

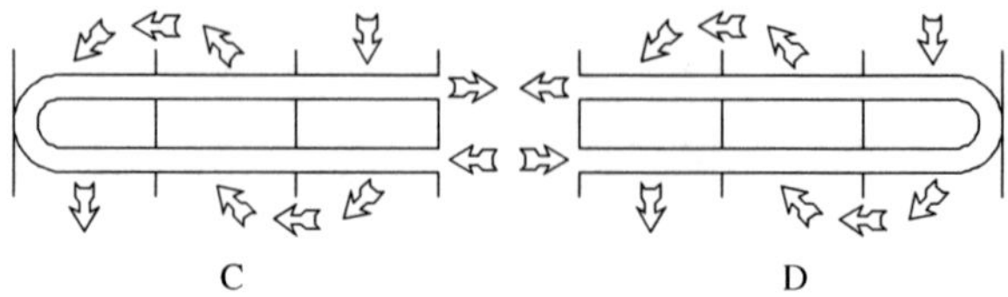

Table A.5 – Heat exchangers – C and D type – 3 passages of external fluid Corrective factor χ_c

β \ ψ	0.04	0.08	0.12	0.16	0.20	0.24	0.28	0.32	0.36	0.40	0.44
0.10	**0.639**	*0.840*	*0.901*	*0.932*	0.950	0.962	0.971	0.977	0.982	0.986	0.989
0.15	....	*0.719*	*0.835*	*0.888*	0.919	0.939	0.954	0.964	0.972	0.977	0.982
0.20	....	**0.562**	*0.751*	*0.836*	*0.884*	0.914	0.934	0.949	0.960	0.969	0.976
0.25	....	....	*0.643*	*0.773*	*0.842*	0.884	0.912	0.933	0.948	0.959	0.968
0.30	....	....	**0.513**	*0.693*	*0.791*	0.850	0.888	0.915	0.934	0.949	0.960
0.35	....	....	....	*0.592*	*0.730*	*0.809*	0.859	0.894	0.919	0.938	0.952
0.40	....	....	....	**0.480**	*0.653*	*0.761*	0.827	0.871	0.902	0.925	0.942
0.50	....	....	....	....	**0.459**	*0.629*	*0.742*	0.813	0.862	0.896	0.921
0.60	....	....	....	....	....	**0.449**	*0.618*	*0.734*	0.809	0.859	0.895
0.70	....	....	....	....	....	....	**0.450**	*0.620*	0.737	0.812	0.863
0.80	....	....	....	....	....	....	....	*0.462*	*0.634*	0.748	0.822
0.90	....	....	....	....	....	....	....	**0.333**	*0.488*	*0.659*	0.767
1.00	....	....	....	....	....	....	....	....	**0.353**	*0.530*	0.693
1.10	....	....	....	....	....	....	....	....	....	**0.388**	*0.586*
1.20	....	....	....	....	....	....	....	....	....	....	*0.445*
1.30	....	....	....	....	....	....	....	....	....	....	**0.333**

If β> 1.3 no solution is possible in this field

Values in italics : 2<γ<4 ; **Values in bold type : 4<γ<6**

Table A.5 – (continued)

	ψ										
β	0.48	0.52	0.56	0.60	0.64	0.68	0.72	0.76	0.80	0.84	0.88
0.2	0.981	0.985	0.988	·······	·······	·······	·······	·······	·······	·······	·······
0.3	0.969	0.976	0.982	0.986	0.989	·······	·······	·······	·······	·······	·······
0.4	0.956	0.966	0.974	0.980	0.985	0.989	·······	·······	·······	·······	·······
0.5	0.940	0.954	0.965	0.974	0.981	0.986	·······	·······	·······	·······	·······
0.6	0.921	0.941	0.956	0.967	0.976	0.983	0.988	·······	·······	·······	·······
0.7	0.899	0.925	0.945	0.959	0.970	0.979	0.985	·······	·······	·······	·······
0.8	0.871	0.906	0.932	0.950	0.964	0.975	0.983	0.988	·······	·······	·······
0.9	0.837	0.884	0.916	0.940	0.958	0.970	0.980	0.986	·······	·······	·······
1.0	0.793	0.856	0.899	0.929	0.950	0.965	0.976	0.985	·······	·······	·······
1.1	0.733	0.821	0.877	0.915	0.941	0.960	0.973	0.982	0.989	·······	·······
1.2	0.650	0.776	0.851	0.899	0.931	0.954	0.969	0.980	0.988	·······	·······
1.3	*0.530*	0.715	0.818	0.880	0.920	0.947	0.965	0.978	0.986	·······	·······
1.4	*0.397*	0.629	0.776	0.857	0.907	0.939	0.961	0.975	0.985	·······	·······
1.5	**0.305**	*0.507*	0.719	0.829	0.892	0.931	0.956	0.972	0.983	·······	·······
1.6	·······	*0.381*	0.638	0.793	0.873	0.921	0.950	0.969	0.982	0.990	·······
1.7	·······	**0.298**	*0.523*	0.745	0.851	0.909	0.944	0.966	0.980	0.989	·······
1.8	·······	·······	*0.397*	0.679	0.823	0.896	0.937	0.962	0.978	0.988	·······
1.9	·······	·······	**0.310**	0.584	0.788	0.880	0.930	0.958	0.976	0.987	·······
2.0	·······	·······	·······	*0.459*	0.742	0.861	0.921	0.954	0.974	0.986	·······
2.1	·······	·······	·······	*0.353*	0.677	0.838	0.911	0.950	0.972	0.985	·······
2.2	·······	·······	·······	**0.282**	0.584	0.810	0.899	0.944	0.969	0.984	·······
2.3	·······	·······	·······	·······	*0.462*	0.773	0.886	0.939	0.967	0.983	·······
2.4	·······	·······	·······	·······	*0.358*	0.723	0.870	0.932	0.964	0.982	·······
2.5	·······	·······	·······	·······	**0.287**	0.653	0.851	0.925	0.961	0.980	·······
2.6	·······	·······	·······	·······	·······	0.552	0.827	0.917	0.958	0.979	·······
2.7	·······	·······	·······	·······	·······	*0.431*	0.797	0.908	0.954	0.978	0.990
2.8	·······	·······	·······	·······	·······	0.339	0.758	0.897	0.950	0.976	0.989
2.9	·······	·······	·······	·······	·······	**0.274**	0.704	0.884	0.946	0.975	0.989
3.0	·······	·······	·······	·······	·······	·······	0.627	0.869	0.941	0.973	0.988

If β<0.2 or ψ >0.88 use the mean logarithmic temperature difference for counter flow
Values in italics : 2<γ<4 ; **Values in bold type : 4<γ<6**

Fig. A.6

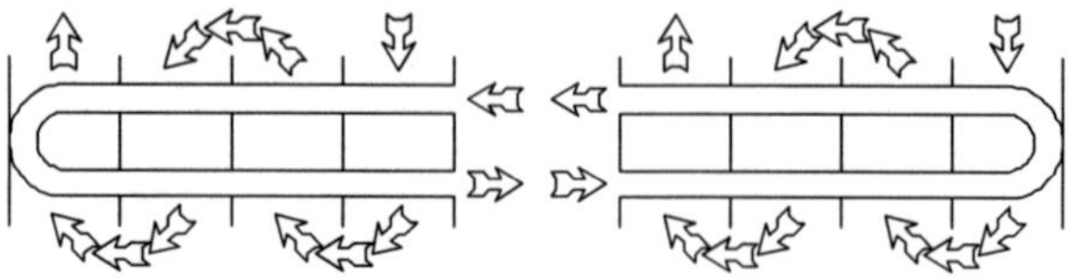

A D

Table A.6 – Heat exchangers – A and D type – 4 passages of external fluid Corrective factor χ_c (valid also for I and N type with 4 passages of external fluid)

β \ ψ	0.04	0.08	0.12	0.16	0.20	0.24	0.28	0.32	0.36	0.40	0.44
0.05	**0.761**	*0.907*	*0.944*	0.962	0.972	0.979	0.984	0.987	0.990	....	....
0.10	....	*0.764*	*0.872*	*0.915*	0.940	0.955	0.965	0.973	0.979	0.983	0.987
0.15	....	....	*0.771*	*0.858*	0.901	0.927	0.945	0.957	0.967	0.974	0.979
0.20	....	....	**0.560**	*0.782*	*0.854*	0.895	0.921	0.940	0.954	0.964	0.972
0.25	....	....	....	*0.666*	*0.795*	0.857	0.894	0.920	0.939	0.953	0.963
0.30	....	....	....	....	*0.714*	*0.810*	0.863	0.898	0.922	0.940	0.954
0.35	....	....	....	....	*0.573*	*0.750*	0.825	0.872	0.904	0.927	0.944
0.40	....	....	....	....	....	*0.665*	0.779	0.842	0.883	0.911	0.932
0.45	....	....	....	....	....	....	*0.720*	0.806	0.859	0.894	0.920
0.50	....	....	....	....	....	....	*0.634*	0.762	0.830	0.875	0.907
0.60	....	....	....	....	....	....	....	*0.623*	0.757	0.828	0.874
0.70	....	....	....	....	....	....	....	....	*0.633*	0.762	0.833
0.80	....	....	....	....	....	....	....	....	....	0.658	0.777
0.90	....	....	....	....	....	....	....	....	....	....	0.694
1.00	....	....	....	....	....	....	....	....	....	....	*0.515*

If β> 1.0 no solution is possible in this field
Values in italics : 2<γ<4 , **Values in bold type : 4<γ<6**

Table A.6 – (continued)

	ψ										
β	0.48	0.52	0.56	0.60	0.64	0.68	0.72	0.76	0.80	0.84	0.88
0.2	0.978	0.983	0.986	0.990	....	....	....	....	....	....	....
0.3	0.964	0.972	0.979	0.984	0.988	....	....	....	....	....	....
0.4	0.948	0.960	0.970	0.977	0.983	0.988	....	....	....	....	....
0.5	0.930	0.947	0.960	0.970	0.978	0.984	0.989	....	....	....	....
0.6	0.907	0.931	0.949	0.962	0.972	0.980	0.986	....	....	....	....
0.7	0.879	0.912	0.935	0.953	0.966	0.976	0.983	0.989	....	....	....
0.8	0.845	0.889	0.920	0.942	0.959	0.971	0.980	0.987	....	....	....
0.9	0.799	0.861	0.902	0.930	0.951	0.966	0.977	0.985	....	....	....
1.0	0.735	0.825	0.880	0.916	0.942	0.960	0.973	0.982	0.989	....	....
1.1	0.629	0.777	0.853	0.900	0.932	0.954	0.969	0.980	0.987	....	....
1.2	....	0.708	0.818	0.880	0.920	0.946	0.964	0.977	0.986	....	....
1.3	....	0.584	0.773	0.856	0.906	0.938	0.960	0.974	0.984	....	....
1.4	....	....	0.706	0.826	0.890	0.929	0.954	0.971	0.983	....	....
1.5	....	....	0.588	0.787	0.871	0.919	0.949	0.968	0.981	0.989	....
1.6	....	....	....	0.732	0.847	0.906	0.942	0.964	0.979	0.988	....
1.7	....	....	....	0.646	0.817	0.892	0.935	0.961	0.977	0.987	....
1.8	....	....	....	....	0.778	0.876	0.927	0.957	0.975	0.986	....
1.9	....	....	....	....	0.723	0.855	0.917	0.952	0.973	0.985	....
2.0	....	....	....	....	0.633	0.830	0.907	0.947	0.970	0.984	....
2.1	....	....	....	....	....	0.799	0.894	0.941	0.968	0.983	....
2.2	....	....	....	....	....	0.755	0.880	0.935	0.965	0.982	....
2.3	....	....	....	....	....	0.691	0.862	0.928	0.962	0.980	....
2.4	....	....	....	....	....	0.567	0.841	0.920	0.958	0.979	....
2.5	....	....	....	....	....	....	0.814	0.911	0.955	0.977	0.990
2.6	....	....	....	....	....	....	0.779	0.901	0.951	0.976	0.989
2.7	....	....	....	....	....	....	0.731	0.889	0.947	0.974	0.988
2.8	....	....	....	....	....	....	0.652	0.875	0.942	0.972	0.988
2.9	....	....	....	....	....	....	....	0.859	0.937	0.971	0.987
3.0	....	....	....	....	....	....	....	0.839	0.931	0.969	0.986

If $\beta<0.2$ or $\psi>0.88$ use the mean logarithmic temperature difference for counter flow

Fig. A.7

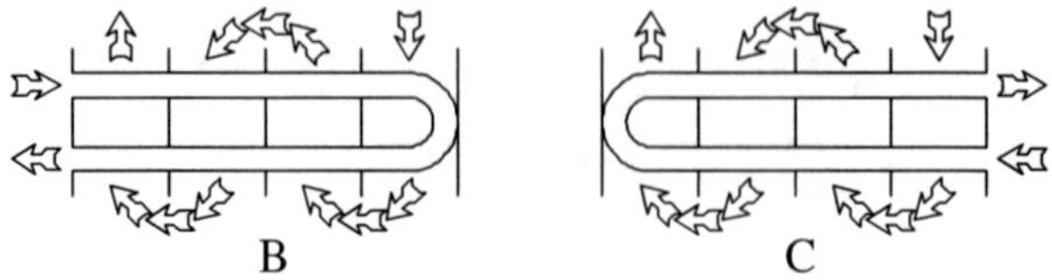

Table A.7 – Heat exchangers – B and C type – 4 passages of external fluid Corrective factor χ_c

	Ψ									
β	0.08	0.12	0.16	0.20	0.24	0.28	0.32	0.36	0.40	0.44
0.10	*0.823*	*0.894*	*0.927*	0.947	0.960	0.969	0.976	0.981	0.985	0.988
0.15	**0.676**	*0.820*	*0.881*	0.914	0.936	0.951	0.962	0.970	0.976	0.982
0.20	....	*0.722*	*0.823*	*0.876*	0.909	0.931	0.946	0.958	0.967	0.974
0.25	....	**0.582**	*0.751*	*0.830*	0.877	0.908	0.929	0.945	0.957	0.967
0.30	....	....	*0.654*	*0.774*	0.839	0.881	0.910	0.931	0.946	0.958
0.35	....	....	**0.520**	*0.702*	*0.795*	0.851	0.888	0.915	0.934	0.949
0.40	....	....	....	*0.607*	*0.740*	0.815	0.863	0.897	0.921	0.940
0.45	....	....	....	**0.481**	*0.670*	*0.772*	0.834	0.877	0.907	0.929
0.50	....	....	....	....	*0.578*	*0.719*	0.801	0.853	0.890	0.917
0.60	....	....	....	....	**0.349**	*0.566*	*0.711*	0.796	0.851	0.889
0.70	....	....	....	....	....	**0.351**	*0.570*	0.714	0.799	0.855
0.80	....	....	....	....	....	....	**0.364**	*0.589*	0.728	0.810
0.90	....	....	....	....	....	....	....	**0.395**	*0.621*	0.750
1.00	....	....	....	....	....	....	....	....	*0.449*	0.663
1.10	....	....	....	....	....	....	....	....	**0.294**	*0.526*
1.20	....	....	....	....	....	....	....	....	....	**0.347**

If β >1.2 or ψ<0.08 no solution is possible in this field
Values in italics : 2<γ <4 ; **Values in bold type : 4<γ<6**

Table A.7 – (continued)

β	ψ 0.48	0.52	0.56	0.60	0.64	0.68	0.72	0.76	0.80	0.84	0.88
0.2	0.980	0.984	0.988	....	....	....	....	....	....	....	....
0.3	0.968	0.975	0.981	0.985	0.989	....	....	....	....	....	....
0.4	0.954	0.964	0.973	0.980	0.985	0.989	....	....	....	....	....
0.5	0.937	0.952	0.964	0.973	0.980	0.986	0.990	....	....	....	....
0.6	0.917	0.938	0.954	0.966	0.975	0.982	0.987	....	....	....	....
0.7	0.894	0.921	0.942	0.958	0.969	0.978	0.985	0.990	....	....	....
0.8	0.864	0.902	0.928	0.948	0.963	0.974	0.982	0.988	....	....	....
0.9	0.827	0.878	0.912	0.937	0.956	0.969	0.979	0.986	....	....	....
1.0	0.778	0.848	0.894	0.925	0.948	0.964	0.976	0.984	0.990	....	....
1.1	0.711	0.810	0.871	0.911	0.939	0.958	0.972	0.982	0.989	....	....
1.2	*0.610*	0.760	0.843	0.894	0.928	0.952	0.968	0.979	0.987	....	....
1.3	*0.449*	0.690	0.807	0.874	0.916	0.944	0.964	0.977	0.986	....	....
1.4	**0.301**	0.582	0.759	0.849	0.902	0.936	0.959	0.974	0.984	....	....
1.5	....	*0.416*	0.693	0.818	0.886	0.927	0.954	0.971	0.983	....	....
1.6	....	**0.288**	0.593	0.778	0.866	0.917	0.948	0.968	0.981	0.989	....
1.7	....	....	*0.435*	0.724	0.842	0.905	0.942	0.964	0.979	0.988	....
1.8	....	....	**0.301**	0.644	0.812	0.890	0.934	0.961	0.977	0.987	....
1.9	....	....	**0.233**	0.517	0.773	0.874	0.926	0.957	0.975	0.986	....
2.0	....	....	....	*0.357*	0.719	0.853	0.917	0.952	0.973	0.986	....
2.1	....	....	....	**0.265**	0.641	0.828	0.906	0.947	0.971	0.985	....
2.2	....	....	....	....	0.515	0.796	0.894	0.942	0.968	0.983	....
2.3	....	....	....	....	*0.359*	0.754	0.880	0.936	0.965	0.982	....
2.4	....	....	....	....	**0.269**	0.696	0.862	0.929	0.963	0.981	....
2.5	....	....	....	....	**0.215**	0.607	0.841	0.921	0.959	0.980	....
2.6	....	....	....	....	....	0.465	0.815	0.913	0.956	0.978	....
2.7	....	....	....	....	....	*0.328*	0.781	0.903	0.952	0.977	0.989
2.8	....	....	....	....	....	**0.255**	0.736	0.891	0.948	0.975	0.989
2.9	....	....	....	....	....	**0.205**	0.671	0.878	0.944	0.973	0.988
3.0	....	....	....	....	....	....	0.568	0.861	0.939	0.972	0.988

If β<0.2 or ψ>0.88 use the mean logarithmic temperature difference for counter flow
Values in italics : 2<γ<4 ; **Values in bold type : 4<γ<6**

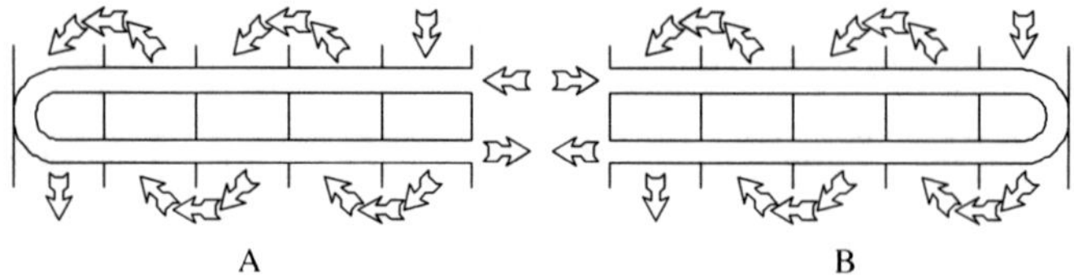

Fig. A.8

Table A.8 – Heat exchangers – A and B type – 5 passages of external fluid Corrective factor χ_c (valid also for I and L type with 5 passages of external fluid)

β \ ψ	0.04	0.08	0.12	0.16	0.20	0.24	0.28	0.32	0.36	0.40	0.44
0.05	**0.778**	*0.910*	*0.946*	0.963	0.972	0.979	0.984	0.987	....	....	....
0.10	....	*0.776*	*0.876*	*0.918*	0.941	0.956	0.966	0.974	0.979	0.984	0.987
0.15	....	....	*0.781*	*0.862*	0.903	0.929	0.946	0.958	0.967	0.974	0.980
0.20	....	....	*0.612*	*0.790*	*0.858*	0.897	0.923	0.941	0.954	0.964	0.972
0.25	....	....	....	*0.685*	*0.802*	0.860	0.897	0.922	0.940	0.953	0.964
0.30	....	....	....	....	*0.726*	*0.815*	0.866	0.900	0.924	0.941	0.954
0.35	....	....	....	....	*0.608*	*0.758*	0.830	0.875	0.906	0.928	0.945.
0.40	....	....	....	....	....	*0.681*	*0.786*	0.846	0.885	0.913	0.934
0.45	....	....	....	....	....	*0.547*	*0.730*	0.811	0.862	0.897	0.922
0.50	....	....	....	....	....	....	*0.653*	0.769	0.835	0.878	0.908
0.60	....	....	....	....	....	....	....	*0.643*	0.764	0.832	0.877
0.70	....	....	....	....	....	....	....	....	*0.651*	0.769	0.837
0.80	....	....	....	....	....	....	....	....	....	0.673	0.784
0.90	....	....	....	....	....	....	....	....	....	....	0.705
1.00	....	....	....	....	....	....	....	....	....	....	*0.560*

If β >1.0 no solution is possible in this field
Values in italics : 2<γ<4 ; **Value in bold type : 4<γ<6**

Table A.8 – (continued)

	ψ										
β	0.48	0.52	0.56	0.60	0.64	0.68	0.72	0.76	0.80	0.84	0.88
0.2	0.978	0.983	0.987	....	....	....	....	....	....	....	....
0.3	0.965	0.973	0.979	0.984	0.988	....	....	....	....	....	....
0.4	0.949	0.961	0.970	0.978	0.984	0.988	....	....	....	....	....
0.5	0.931	0.948	0.961	0.971	0.978	0.984	0.989	....	....	....	....
0.6	0.909	0.932	0.950	0.963	0.973	0.980	0.986	....	....	....	....
0.7	0.882	0.914	0.936	0.954	0.966	0.976	0.983	0.989	....	....	....
0.8	0.848	0.891	0.921	0.943	0.959	0.971	0.980	0.987	....	....	....
0.9	0.804	0.864	0.904	0.932	0.952	0.966	0.977	0.985	....	....	....
1.0	0.743	0.829	0.882	0.918	0.943	0.961	0.973	0.983	0.989	....	....
1.1	0.648	0.783	0.856	0.902	0.933	0.954	0.969	0.980	0.988	....	....
1.2	....	0.718	0.823	0.883	0.921	0.947	0.965	0.978	0.986	....	....
1.3	....	0.610	0.770	0.860	0.908	0.939	0.960	0.975	0.985	....	....
1.4	....	....	0.717	0.831	0.892	0.930	0.955	0.972	0.983	....	....
1.5	....	....	0.614	0.793	0.874	0.920	0.950	0.969	0.981	0.989	....
1.6	....	....	....	0.741	0.851	0.908	0.943	0.965	0.979	0.989	....
1.7	....	....	....	0.663	0.822	0.895	0.936	0.961	0.977	0.988	....
1.8	....	....	....	*0.470*	0.785	0.878	0.928	0.957	0.975	0.986	....
1.9	....	....	....	....	0.733	0.859	0.919	0.953	0.973	0.986	....
2.0	....	....	....	....	0.652	0.835	0.908	0.948	0.971	0.984	....
2.1	....	....	....	....	....	0.804	0.896	0.942	0.968	0.983	....
2.2	....	....	....	....	....	0.763	0.882	0.936	0.965	0.982	....
2.3	....	....	....	....	....	0.705	0.865	0.929	0.962	0.981	....
2.4	....	....	....	....	....	0.603	0.845	0.922	0.959	0.979	....
2.5	....	....	....	....	....	....	0.819	0.913	0.956	0.978	0.990
2.6	....	....	....	....	....	....	0.786	0.903	0.952	0.976	0.989
2.7	....	....	....	....	....	....	0.741	0.892	0.948	0.975	0.989
2.8	....	....	....	....	....	....	0.671	0.878	0.943	0.973	0.988
2.9	....	....	....	....	....	....	0.512	0.862	0.938	0.971	0.987
3.0	....	....	....	....	....	....	....	0.843	0.933	0.969	0.986

If $\beta<0.2$ or $\psi > 0.88$ use the mean logarithmic temperature difference for counter flow
Value in italics : $2<\gamma<4$

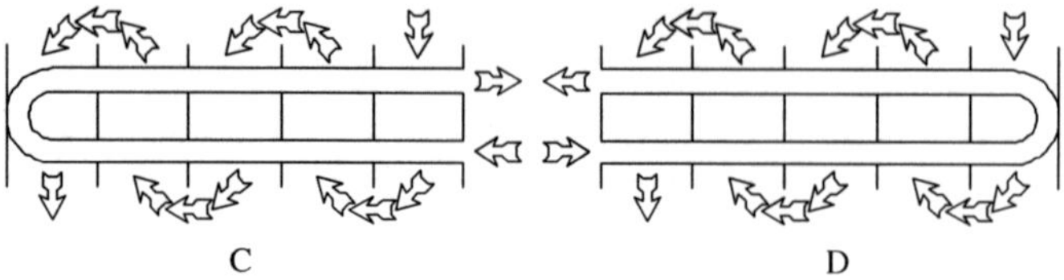

Fig. A.9

Table A.9 – Heat exchangers – C and D type – 5 passages of external fluid Corrective factor χ_c

β \ ψ	0.08	0.12	0.16	0.20	0.24	0.28	0.32	0.36	0.40	0.44
0.10	*0.814*	*0.890*	*0.925*	0.946	0.959	0.969	0.975	0.981	0.985	0.988
0.15	**0.648**	*0.813*	*0.877*	0.912	0.935	0.950	0.961	0.970	0.976	0.981
0.20	....	*0.706*	*0.817*	*0.872*	0.906	0.929	0.945	0.958	0.967	0.974
0.25	....	**0.538**	*0.739*	*0.824*	0.873	0.905	0.928	0.944	0.957	0.966
0.30	....	....	*0.631*	*0,765*	0.835	0.878	0.908	0.929	0.945	0.958
0.35	....	....	**0.464**	*0.687*	*0.788*	0.847	0.885	0.913	0.933	0.948
0.40	....	....	....	*0.578*	*0.729*	0.809	0.859	0.894	0.919	0.938
0.45	....	....	....	**0.418**	*0.652*	*0.764*	0.830	0.874	0.904	0.927
0.50	....	....	....	....	*0.545*	*0.707*	0.794	0.849	0.888	0.915
0.60	....	....	....	....	....	*0.531*	*0.699*	0.789	0.847	0.887
0.70	....	....	....	....	....	....	*0.537*	*0.703*	0.793	0.851
0.80	....	....	....	....	....	....	**0.296**	*0.561*	0.718	0.805
0.90	....	....	....	....	....	....	....	**0.326**	*0.599*	0.741
1.00	....	....	....	....	....	....	....	....	**0.386**	0.647
1.10	....	....	....	....	....	....	....	....	....	*0.483*
1.20	....	....	....	....	....	....	....	....	....	**0.280**

If β>1.2 or ψ<0.08 no solution is possible in this field
Values in italics : 2<γ<4 ; **Values in bold type : 4<γ<6**

Table A.9 – (continued)

	ψ										
β	0.48	0.52	0.56	0.60	0.64	0.68	0.72	0.76	0.80	0.84	0.88
0.2	0.980	0.984	0.988	....	....	....	....	....	....	....	....
0.3	0.967	0.975	0.980	0.985	0.989	....	....	....	....	....	....
0.4	0.953	0.964	0.972	0.979	0.985	0.989	....	....	....	....	....
0.5	0.936	0.951	0.963	0.973	0.980	0.985	0.990	....	....	....	....
0.6	0.915	0.937	0.953	0.965	0.975	0.982	0.987	....	....	....	....
0.7	0.891	0.920	0.941	0.957	0.969	0.978	0.985	0.990	....	....	....
0.8	0.861	0.899	0.927	0.947	0.962	0.973	0.982	0.988	....	....	....
0.9	0.822	0.875	0.911	0.936	0.955	0.968	0.978	0.986	....	....	....
1.0	0.771	0.844	0.891	0.924	0.947	0.963	0.975	0.984	0.990	....	....
1.1	0.699	0.805	0.868	0.909	0.937	0.957	0.971	0.981	0.988	....	....
1.2	*0.586*	0.752	0.838	0.892	0.927	0.951	0.967	0.979	0.987	....	....
1.3	*0.385*	0.676	0.801	0.871	0.914	0.943	0.963	0.976	0.986	....	....
1.4	**0.241**	*0.552*	0.751	0.845	0.900	0.935	0.958	0.974	0.984	....	....
1.5	....	*0.347*	0.680	0.813	0.883	0.926	0.953	0.971	0.982	....	....
1.6	....	**0.230**	0.565	0.771	0.863	0.915	0.947	0.967	0.981	0.989	....
1.7	....	....	*0.367*	0.712	0.838	0.902	0.940	0.964	0.979	0.988	....
1.8	....	....	**0.241**	0.624	0.806	0.888	0.933	0.960	0.977	0.987	....
1.9	....	....	....	*0.468*	0.765	0.871	0.925	0.956	0.975	0.986	....
2.0	....	....	....	*0.288*	0.707	0.849	0.915	0.951	0.973	0.985	....
2.1	....	....	....	**0.212**	0.619	0.823	0.904	0.946	0.970	0.984	....
2.2	....	....	....	....	*0.463*	0.790	0.892	0.941	0.968	0.983	....
2.3	....	....	....	....	*0.290*	0.745	0.877	0.934	0.965	0.982	....
2.4	....	....	....	....	**0.215**	0.681	0.859	0.927	0.962	0.981	....
2.5	....	....	....	....	....	0.577	0.837	0.920	0.958	0.979	....
2.6	....	....	....	....	....	*0.396*	0.809	0.911	0.955	0.978	0.990
2.7	....	....	....	....	....	*0.263*	0.774	0.900	0.951	0.976	0.989
2.8	....	....	....	....	....	**0.204**	0.725	0.888	0.947	0.975	0.989
2.9	....	....	....	....	....	....	0.651	0.874	0.942	0.973	0.988
3.0	....	....	....	....	....	....	0.526	0.858	0.937	0.971	0.987

If β<0.2 or ψ>0.88 use the mean logarithmic temperature difference for counter flow
Values in italics : 2<γ<4 ; **Values in bold type : 4<γ<6**

A.2.2 Heat Exchangers with 3 Passages of Internal Fluid (Fig. 2.6)

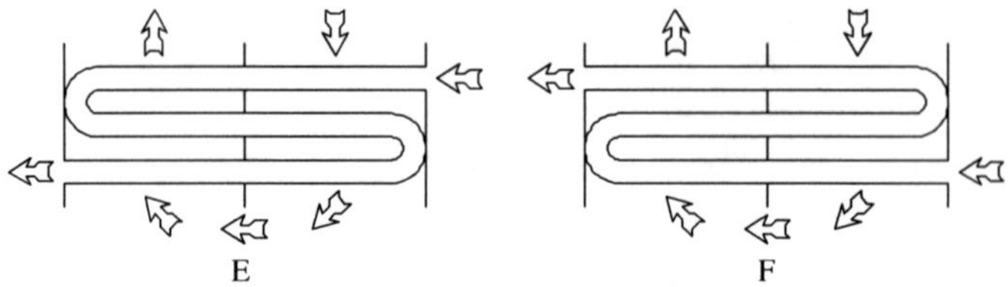

Fig. A.10

Table A.10 – Heat exchangers – E and F type – 2 passages of external fluid Corrective factor χ_c

β \ ψ	0.04	0.08	0.12	0.16	0.20	0.24	0.28	0.32	0.36	0.40	0.44
0.05	**0.723**	*0.901*	*0.941*	0.960	0.971	0.978	0.983	0.987	0.989	....	....
0.10	....	*0.736*	*0.864*	*0.911*	0.937	0.953	0.964	0.972	0.978	0.983	0.986
0.15	....	....	*0.750*	*0.849*	0.896	0.924	0.942	0.956	0.965	0.973	0.979
0.20	....	....	....	*0.765*	*0.846*	0.890	0.918	0.937	0.952	0.962	0.970
0.25	....	....	....	*0.616*	*0.781*	0.849	0.889	0.917	0.936	0.950	0.962
0.30	....	....	....	....	*0.686*	*0.798*	0.856	0.893	0.919	0.938	0.952
0.35	....	....	....	....	....	*0.730*	0.815	0.866	0.899	0.923	0.941
0.40	....	....	....	....	....	*0.624*	*0.765*	0.833	0.877	0.908	0.930
0.45	....	....	....	....	....	....	*0.696*	0.795	0.851	0.890	0.917
0.50	....	....	....	....	....	....	*0.582*	0.746	0.821	0.869	0.903
0.60	....	....	....	....	....	....	....	*0.569*	0.740	0.819	0.869
0.70	....	....	....	....	....	....	....	....	*0.586*	0.747	0.825
0.80	....	....	....	....	....	....	....	....	....	*0.623*	0.764
0.90	....	....	....	....	....	....	....	....	....	....	0.668

If β>0.9 no solution is possible in this field

Values in italics : 2<γ<4 ; **Value in bold type : 4<γ<6**

Table A.10 – (continued)

β \ ψ	0.48	0.52	0.56	0.60	0.64	0.68	0.72	0.76	0.80	0.84	0.88
0.1	0.989	....	....	....	....	....	....	....	....	....	....
0.2	0.977	0.982	0.986	0.989	....	....	....	....	....	....	....
0.3	0.963	0.971	0.978	0.983	0.987	....	....	....	....	....	....
0.4	0.946	0.959	0.969	0.976	0.983	0.987	....	....	....	....	....
0.5	0.927	0.945	0.958	0.969	0.977	0.984	0.988	....	....	....	....
0.6	0.903	0.928	0.947	0.961	0.971	0.979	0.986	....	....	....	....
0.7	0.874	0.908	0.933	0.951	0.965	0.975	0.983	0.988	....	....	....
0.8	0.837	0.884	0.917	0.940	0.957	0.970	0.979	0.986	....	....	....
0.9	0.788	0.854	0.898	0.928	0.949	0.964	0.976	0.984	0.990	....	....
1.0	0.716	0.816	0.874	0.913	0.940	0.958	0.972	0.982	0.988	....	....
1.1	0.583	0.764	0.846	0.896	0.929	0.952	0.968	0.979	0.987	....	....
1.2	....	0.685	0.809	0.875	0.916	0.944	0.963	0.976	0.986	....	....
1.3	....	*0.489*	0.758	0.849	0.902	0.936	0.958	0.973	0.984	....	....
1.4	....	....	0.683	0.817	0.885	0.926	0.953	0.970	0.982	0.990	....
1.5	....	....	*0.499*	0.774	0.865	0.915	0.946	0.967	0.980	0.989	....
1.6	....	....	....	0.713	0.839	0.902	0.940	0.963	0.978	0.988	....
1.7	....	....	....	0.604	0.807	0.888	0.932	0.959	0.976	0.987	....
1.8	....	....	....	....	0.764	0.870	0.924	0.955	0.974	0.986	....
1.9	....	....	....	....	0.701	0.848	0.914	0.950	0.972	0.985	....
2.0	....	....	....	....	0.583	0.821	0.902	0.945	0.969	0.984	....
2.1	....	....	....	....	....	0.786	0.889	0.939	0.966	0.982	....
2.2	....	....	....	....	....	0.738	0.874	0.932	0.963	0.981	....
2.3	....	....	....	....	....	0.662	0.855	0.925	0.960	0.980	....
2.4	....	....	....	....	....	....	0.833	0.917	0.957	0.978	0.990
2.5	....	....	....	....	....	....	0.804	0.908	0.953	0.976	0.989
2.6	....	....	....	....	....	....	0.765	0.897	0.949	0.975	0.989
2.7	....	....	....	....	....	....	0.709	0.884	0.945	0.973	0.988
2.8	....	....	....	....	....	....	0.607	0.870	0.940	0.971	0.987
2.9	....	....	....	....	....	....	....	0.852	0.934	0.969	0.987
3.0	....	....	....	....	....	....	....	0.830	0.928	0.967	0.986

If $\psi > 0.88$ use the mean logarithmic temperature difference for counter flow
Values in italics : $2 < \gamma < 4$

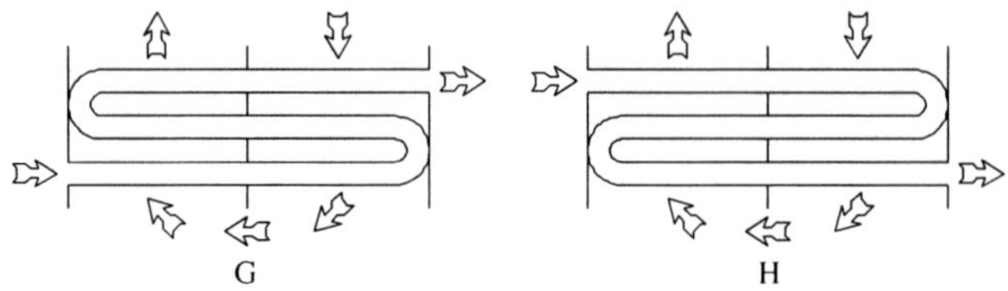

Fig. A.11

Table A.11 – Heat exchangers – G and H type – 2 passages of external fluid Corrective factor χ_c

β \ ψ	0.04	0.08	0.12	0.16	0.20	0.24	0.28	0.32	0.36	0.40	0.44
0.10	**0.640**	*0.840*	*0.901*	*0.932*	0.950	0.962	0.971	0.977	0.982	0.986	0.989
0.15	....	*0.720*	*0.835*	*0.888*	0.919	0.939	0.954	0.964	0.972	0.977	0.982
0.20	....	**0.562**	*0.751*	*0.836*	*0.884*	0.914	0.934	0.949	0.960	0.969	0.975
0.25	....	....	*0.643*	*0.773*	*0.842*	0.884	0.912	0.933	0.948	0.959	0.968
0.30	....	....	**0.513**	*0.693*	*0.792*	0.850	0.888	0.915	0.934	0.949	0.960
0.35	....	....	....	*0.593*	*0.730*	*0.809*	0.859	0.894	0.919	0.938	0.952
0.40	....	....	....	**0.481**	*0.654*	*0.761*	0.827	0.871	0.902	0.925	0.942
0.50	....	....	....	....	**0.461**	*0.630*	*0.742*	0.813	0.862	0.896	0.921
0.60	....	....	....	....	....	**0.452**	*0.619*	*0.735*	0.808	0.859	0.895
0.70	....	....	....	....	....	....	**0.453**	*0.621*	0.737	0.812	0.863
0.80	....	....	....	....	....	....	....	*0.465*	*0.634*	0.749	0.822
0.90	....	....	....	....	....	....	....	**0.340**	*0.491*	*0.660*	0.768
1.00	....	....	....	....	....	....	....	....	**0.361**	*0.533*	0.694
1.10	....	....	....	....	....	....	....	....	....	**0.395**	*0.588*
1.20	....	....	....	....	....	....	....	....	....	**0.306**	*0.451*
1.30	....	....	....	....	....	....	....	....	....	....	**0.344**

If $\beta > 1.3$ no solution is possible in this field

Values in italics : $2<\gamma<4$; **Values in bold type : $4<\gamma<6$**

Table A.11 – (continued)

	ψ										
β	0.48	0.52	0.56	0.60	0.64	0.68	0.72	0.76	0.80	0.84	0.88
0.2	0.981	0.985	0.988	....	....	....	....	....	....	....	....
0.3	0.969	0.976	0.982	0.986	0.989	....	....	....	....	....	....
0.4	0.956	0.966	0.974	0.980	0.985	0.989	....	....	....	....	....
0.5	0.940	0.954	0.965	0.974	0.981	0.986	....	....	....	....	....
0.6	0.921	0.941	0.956	0.967	0.976	0.983	0.988	....	....	....	....
0.7	0.899	0.925	0.945	0.959	0.970	0.979	0.985	....	....	....	....
0.8	0.871	0.906	0.932	0.950	0.964	0.975	0.983	0.988	....	....	....
0.9	0.837	0.884	0.917	0.940	0.958	0.970	0.980	0.986	....	....	....
1.0	0.793	0.856	0.899	0.929	0.950	0.965	0.976	0.984	....	....	....
1.1	0.734	0.821	0.877	0.915	0.941	0.960	0.973	0.982	0.989	....	....
1.2	0.651	0.776	0.851	0.899	0.931	0.954	0.969	0.980	0.988	....	....
1.3	*0.534*	0.716	0.819	0.880	0.920	0.947	0.965	0.978	0.986	....	....
1.4	*0.406*	0.631	0.776	0.858	0.907	0.939	0.961	0.975	0.985	....	....
1.5	**0.319**	*0.512*	0.720	0.829	0.892	0.931	0.956	0.972	0.983	....	....
1.6	....	*0.392*	0.640	0.793	0.874	0.921	0.950	0.969	0.982	0.990	....
1.7	....	**0.314**	*0.528*	0.745	0.851	0.909	0.944	0.966	0.980	0.989	....
1.8	....	....	*0.409*	0.680	0.824	0.896	0.937	0.963	0.978	0.988	....
1.9	....	....	**0.327**	0.588	0.789	0.880	0.929	0.958	0.976	0.987	....
2.0	....	....	**0.271**	*0.469*	0.743	0.862	0.921	0.954	0.974	0.986	....
2.1	....	....	....	*0.369*	0.679	0.839	0.911	0.950	0.972	0.985	....
2.2	....	....	....	**0.304**	0.589	0.810	0.899	0.944	0.969	0.984	....
2.3	....	....	....	....	*0.473*	0.773	0.886	0.939	0.967	0.983	....
2.4	....	....	....	....	*0.376*	0.724	0.870	0.932	0.964	0.982	....
2.5	....	....	....	....	*0.310*	0.656	0.851	0.925	0.961	0.980	....
2.6	....	....	....	....	**0.258**	0.558	0.827	0.917	0.958	0.979	....
2.7	....	....	....	....	....	*0.446*	0.798	0.908	0.954	0.978	....
2.8	....	....	....	....	....	*0.360*	0.759	0.897	0.950	0.976	0.989
2.9	....	....	....	....	....	*0.301*	0.706	0.884	0.946	0.975	0.989
3.0	....	....	....	....	....	**0.268**	0.632	0.870	0.942	0.973	0.988

If $\beta < 0.2$ or $\psi > 0.88$ use the mean logarithmic temperature difference for counter flow
Values in italics : $2<\gamma<4$; **Values in bold type : $4<\gamma<6$**

Fig. A.12

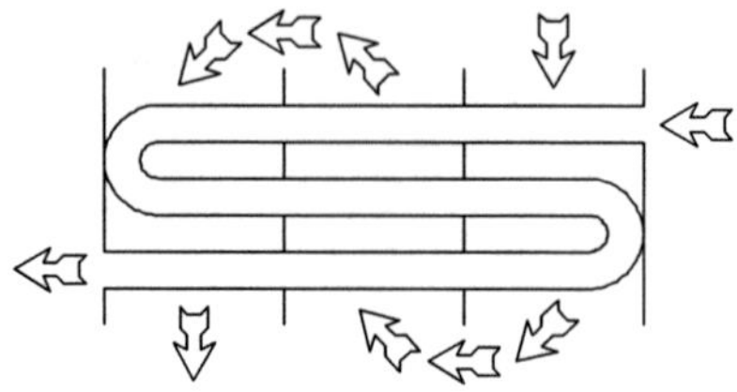

E

Table A.12 – Heat exchangers – E type – 3 passages of external fluid Corrective factor χ_c

β \ ψ	0.04	0.08	0.12	0.16	0.20	0.24	0.28	0.32	0.36	0.40	0.44
0.05	**0.608**	*0.887*	*0.934*	0.955	0.967	0.975	0.981	0.985	0.988	....	....
0.10	....	**0.668**	*0.844*	*0.899*	0.929	0.947	0.960	0.969	0.976	0.981	0.985
0.15	....	....	*0.696*	*0.827*	0.882	0.914	0.936	0.950	0.962	0.970	0.976
0.20	....	....	....	*0.720*	*0.823*	0.875	0.908	0.930	0.946	0.958	0.967
0.25	....	....	....	....	*0.743*	0.827	0.875	0.906	0.929	0.945	0.957
0.30	....	....	....	....	*0.602*	*0.765*	0.836	0.879	0.909	0.930	0.946
0.35	....	....	....	....	....	*0.676*	0.788	0.848	0.887	0.914	0.934
0.40	....	....	....	....	....	....	*0.724*	0.810	0.861	0.896	0.921
0.45	....	....	....	....	....	....	*0.627*	0.763	0.831	0.876	0.907
0.50	......	....	....	....	....	....	....	*0.700*	0.796	0.852	0.891
0.60	....	....	....	....	....	....	....	....	0.693	0.793	0.852
0.70	....	....	....	....	....	....	....	....	....	0.703	0.800
0.80	....	....	....	....	....	....	....	....	....	....	0.725
0.90	....	....	....	....	....	....	....	....	....	....	*0.584*

If β> 0.9 no solution is possible in this field
Values in italics : 2<γ<4 ; **Values in bold type : 4<γ<6**

Table A.12 – (continued)

β	ψ 0.48	0.52	0.56	0.60	0.64	0.68	0.72	0.76	0.80	0.84	0.88
0.1	0.988	....	....	....	....	....	....	....	....	....	....
0.2	0.974	0.980	0.985	0.988	....	....	....	....	....	....	....
0.3	0.959	0.968	0.976	0.982	0.986	0.990	....	....	....	....	....
0.4	0.940	0.954	0.965	0.974	0.981	0.986	....	....	....	....	....
0.5	0.918	0.939	0.954	0.966	0.975	0.982	0.987	....	....	....	....
0.6	0.891	0.920	0.940	0.956	0.968	0.977	0.984	0.989	....	....	....
0.7	0.858	0.897	0.925	0.946	0.961	0.972	0.981	0.987	....	....	....
0.8	0.815	0.870	0.907	0.933	0.953	0.967	0.977	0.985	....	....	....
0.9	0.755	0.835	0.885	0.919	0.943	0.960	0.973	0.982	0.989	....	....
1.0	0.661	0.789	0.858	0.902	0.933	0.954	0.969	0.980	0.987	....	....
1.1	....	0.725	0.825	0.883	0.921	0.946	0.964	0.977	0.986	....	....
1.2	....	0.614	0.781	0.859	0.906	0.938	0.959	0.974	0.984	....	....
1.3	....	....	0.718	0.829	0.890	0.928	0.954	0.970	0.982	0.990	....
1.4	....	....	0.609	0.791	0.871	0.918	0.947	0.967	0.980	0.989	....
1.5	....	....	....	0.738	0.847	0.905	0.940	0.963	0.978	0.988	....
1.6	....	....	....	0.655	0.817	0.891	0.933	0.959	0.976	0.987	....
1.7	....	....	....	....	0.778	0.874	0.924	0.955	0.974	0.986	....
1.8	....	....	....	....	0.725	0.853	0.914	0.950	0.971	0.984	....
1.9	....	....	....	....	0.636	0.828	0.903	0.944	0.969	0.983	....
2.0	....	....	....	....	....	0.796	0.891	0.938	0.966	0.982	....
2.1	....	....	....	....	....	0.752	0.875	0.932	0.963	0.980	....
2.2	....	....	....	....	....	0.689	0.858	0.924	0.959	0.979	....
2.3	....	....	....	....	....	0.561	0.836	0.916	0.956	0.977	0.989
2.4	....	....	....	....	....	....	0.809	0.907	0.952	0.976	0.989
2.5	....	....	....	....	....	....	0.773	0.896	0.948	0.974	0.988
2.6	....	....	....	....	....	....	0.724	0.884	0.943	0.972	0.987
2.7	....	....	....	....	....	....	0.644	0.869	0.938	0.970	0.987
2.8	....	....	....	....	....	....	....	0.852	0.933	0.968	0.986
2.9	....	....	....	....	....	....	....	0.831	0.927	0.966	0.985
3.0	....	....	....	....	....	....	....	0.805	0.920	0.964	0.984

If $\psi > 0.88$ use the mean logarithmic temperature difference for counter flow

Fig. A.13

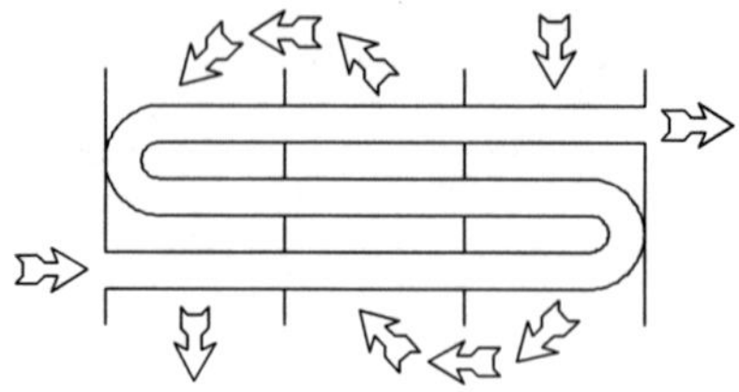

G

Table A.13 – Heat exchangers – G type – 3 passages of external fluid Corrective factor χ_c

β \ ψ	0.04	0.08	0.12	0.16	0.20	0.24	0.28	0.32	0.36	0.40	0.44
0.1	**0.740**	*0.871*	*0.918*	*0.942*	0.957	0.967	0.975	0.980	0.984	0.987	....
0.2	....	**0.677**	*0.802*	*0.865*	0.902	0.926	0.943	0.956	0.966	0.973	0.979
0.3	....	....	**0.637**	*0.757*	*0.828*	0.873	0.905	0.927	0.943	0.956	0.966
0.4	....	....	**0.469**	*0.609*	*0.726*	*0.803*	0.855	0.890	0.916	0.935	0.950
0.5	....	....	....	**0.461**	*0.590*	*0.707*	*0.788*	0.843	0.882	0.911	0.932
0.6	....	....	....	....	**0.456**	*0.581*	*0.698*	0.782	0.839	0.880	0.910
0.7	....	....	....	....	....	**0.456**	*0.580*	*0.699*	0.783	0.842	0.883
0.8	....	....	....	....	....	....	**0.460**	*0.589*	*0.708*	0.792	0.850
0.9	....	....	....	....	....	....	....	**0.472**	*0.607*	0.726	0.807
1.0	....	....	....	....	....	....	....	**0.387**	*0.493*	*0.636*	0.750
1.1	....	....	....	....	....	....	....	....	**0.403**	*0.527*	*0.675*
1.2	....	....	....	....	....	....	....	....	....	**0.429**	*0.576*
1.3	....	....	....	....	....	....	....	....	....	**0.363**	*0.471*
1.4	....	....	....	....	....	....	....	....	....	....	**0.392**
1.5	....	....	....	....	....	....	....	....	....	....	**0.337**

If $\beta > 1.5$ no solution is possible in this field

Values in italics : $2<\gamma<4$; **Values in bold type : $4<\gamma<6$**

Table A.13 – (continued)

	ψ									
β	0.48	0.52	0.56	0.60	0.64	0.68	0.72	0.76	0.80	0.84
0.2	0.983	0.987	....	....	....	....	....	....	....	....
0.3	0.973	0.979	0.984	0.988	....	....	....	....	....	....
0.4	0.961	0.970	0.977	0.983	0.987	....	....	....	....	....
0.5	0.948	0.960	0.970	0.977	0.983	0.988	....	....	....	....
0.6	0.932	0.949	0.962	0.971	0.979	0.985	0.989	....	....	....
0.7	0.913	0.935	0.952	0.964	0.974	0.982	0.987	....	....	....
0.8	0.890	0.919	0.941	0.957	0.969	0.978	0.985	....	....	....
0.9	0.862	0.900	0.928	0.948	0.963	0.974	0.982	0.988	....	....
1.0	0.826	0.878	0.913	0.938	0.957	0.970	0.979	0.986	....	....
1.1	0.781	0.849	0.895	0.927	0.949	0.965	0.976	0.985	....	....
1.2	0.719	0.813	0.874	0.913	0.941	0.960	0.973	0.983	0.989	....
1.3	*0.638*	0.767	0.847	0.897	0.931	0.954	0.970	0.981	0.988	....
1.4	*0.536*	0.705	0.813	0.879	0.920	0.947	0.966	0.978	0.987	....
1.5	*0.440*	0.623	0.770	0.856	0.907	0.940	0.961	0.976	0.985	....
1.6	**0.372**	*0.522*	0.712	0.827	0.892	0.931	0.957	0.973	0.984	....
1.7	....	*0.432*	0.635	0.790	0.874	0.922	0.951	0.970	0.983	....
1.8	....	**0.367**	*0.538*	0.742	0.851	0.911	0.946	0.967	0.981	0.990
1.9	....	....	*0.446*	0.677	0.824	0.898	0.939	0.964	0.979	0.989
2.0	....	....	**0.378**	0.591	0.789	0.882	0.932	0.960	0.977	0.988
2.1	....	....	**0.325**	*0.494*	0.742	0.864	0.923	0.956	0.975	0.987
2.2	....	....	....	*0.414*	0.679	0.841	0.914	0.952	0.973	0.986
2.3	....	....	....	**0.354**	0.596	0.812	0.903	0.947	0.971	0.985
2.4	....	....	....	....	*0.500*	0.776	0.890	0.942	0.969	0.984
2.5	....	....	....	....	*0.419*	0.727	0.874	0.935	0.966	0.983
2.6	....	....	....	....	**0.358**	0.660	0.855	0.928	0.963	0.982
2.7	....	....	....	....	....	0.573	0.832	0.921	0.960	0.981
2.8	....	....	....	....	....	*0.480*	0.802	0.912	0.957	0.979
2.9	....	....	....	....	....	*0.404*	0.764	0.901	0.953	0.978
3.0	....	....	....	....	....	**0.343**	0.712	0.889	0.949	0.976

If $\beta < 0.2$ or $\psi > 0.84$ use the mean logarithmic temperature difference
Values in italics : $2<\gamma<4$; **Values in bold type : $4<\gamma<6$**

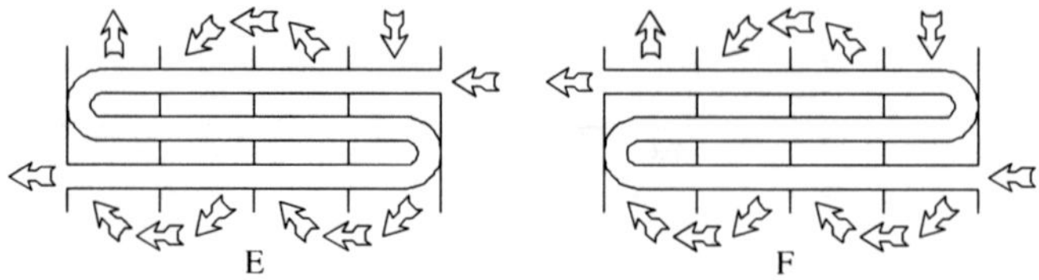

Fig. A.14

Table A.14 – Heat exchangers – E and F type – 4 passages of external fluid Corrective factor χ_c

β \ ψ	0.04	0.08	0.12	0.16	0.20	0.24	0.28	0.32	0.36	0.40	0.44
0.05	**0.764**	*0.904*	*0.942*	0.960	0.971	0.978	0.983	0.986	0.989	....	....
0.10	....	*0.758*	*0.867*	*0.912*	0.937	0.953	0.964	0.972	0.978	0.982	0.986
0.15	....	....	*0.761*	*0.851*	0.896	0.923	0.942	0.955	0.965	0.973	0.978
0.20	....	....	**0.546**	*0.771*	*0.847*	0.889	0.917	0.937	0.951	0.962	0.970
0.25	....	....	....	*0.645*	*0.783*	0.848	0.889	0.916	0.935	0.950	0.961
0.30	....	....	....	....	*0.694*	*0.798*	0.855	0.892	0.918	0.937	0.951
0.35	....	....	....	....	*0.520*	*0.732*	0.814	0.864	0.898	0.922	0.940
0.40	....	....	....	....	....	*0.634*	*0.764*	0.832	0.875	0.906	0.928
0.45	....	....	....	....	....	....	*0.696*	0.792	0.849	0.888	0.915
0.50	....	....	....	....	....	....	*0.590*	0.743	0.819	0.867	0.901
0.60	....	....	....	....	....	....	....	*0.570*	0.736	0.816	0.866
0.70	....	....	....	....	....	....	....	....	*0.579*	0.742	0.821
0.80	....	....	....	....	....	....	....	....	....	*0.612*	0.758
0.90	....	....	....	....	....	....	....	....	....	....	0.657

If $\beta > 0.9$ no solution is possible in this field
Values in italics : $2<\gamma<4$; **Value in bold type : $4<\gamma<6$**

Table A.14 – (continued)

β \ ψ	0.48	0.52	0.56	0.60	0.64	0.68	0.72	0.76	0.80	0.84	0.88
0.1	0.989	…….	…….	…….	…….	…….	…….	…….	…….	…….	…….
0.2	0.976	0.982	0.986	0.989	…….	…….	…….	…….	…….	…….	…….
0.3	0.962	0.971	0.978	0.983	0.987	…….	…….	…….	…….	…….	…….
0.4	0.945	0.958	0.968	0.976	0.982	0.987	…….	…….	…….	…….	…….
0.5	0.925	0.944	0.958	0.969	0.977	0.983	0.988	…….	…….	…….	…….
0.6	0.901	0.927	0.946	0.960	0.971	0.979	0.985	…….	…….	…….	…….
0.7	0.871	0.906	0.931	0.950	0.964	0.974	0.982	0.988	…….	…….	…….
0.8	0.833	0.882	0.915	0.939	0.956	0.969	0.979	0.986	…….	…….	…….
0.9	0.782	0.851	0.895	0.926	0.948	0.964	0.975	0.984	0.990	…….	…….
1.0	0.707	0.811	0.871	0.911	0.938	0.958	0.971	0.981	0.988	…….	…….
1.1	*0.557*	0.757	0.842	0.893	0.927	0.951	0.967	0.979	0.987	…….	…….
1.2	…….	0.672	0.803	0.872	0.915	0.943	0.963	0.976	0.985	…….	…….
1.3	…….	…….	0.750	0.845	0.900	0.934	0.957	0.973	0.984	…….	…….
1.4	…….	…….	0.668	0.812	0.882	0.924	0.952	0.970	0.982	0.990	…….
1.5	…….	…….	…….	0.766	0.861	0.913	0.945	0.966	0.980	0.989	…….
1.6	…….	…….	…….	0.700	0.835	0.900	0.938	0.962	0.978	0.988	…….
1.7	…….	…….	…….	0.569	0.801	0.885	0.931	0.958	0.976	0.987	…….
1.8	…….	…….	…….	…….	0.755	0.866	0.922	0.954	0.973	0.986	…….
1.9	…….	…….	…….	…….	0.686	0.844	0.912	0.949	0.971	0.984	…….
2.0	…….	…….	…….	…….	0.527	0.815	0.900	0.943	0.968	0.983	…….
2.1	…….	…….	…….	…….	…….	0.778	0.886	0.937	0.966	0.982	…….
2.2	…….	…….	…….	…….	…….	0.726	0.870	0.931	0.963	0.981	…….
2.3	…….	…….	…….	…….	…….	0.638	0.851	0.923	0.959	0.979	…….
2.4	…….	…….	…….	…….	…….	…….	0.827	0.915	0.956	0.978	0.990
2.5	…….	…….	…….	…….	…….	…….	0.796	0.905	0.952	0.976	0.989
2.6	…….	…….	…….	…….	…….	…….	0.754	0.894	0.948	0.974	0.988
2.7	…….	…….	…….	…….	…….	…….	0.692	0.881	0.943	0.973	0.988
2.8	…….	…….	…….	…….	…….	…….	0.556	0.865	0.938	0.971	0.987
2.9	…….	…….	…….	…….	…….	…….	…….	0.847	0.933	0.969	0.986
3.0	…….	…….	…….	…….	…….	…….	…….	0.824	0.926	0.967	0.986

If $\psi > 0.88$ use the mean logarithmic temperature difference for counter flow
Value in italics : $2<\gamma<4$

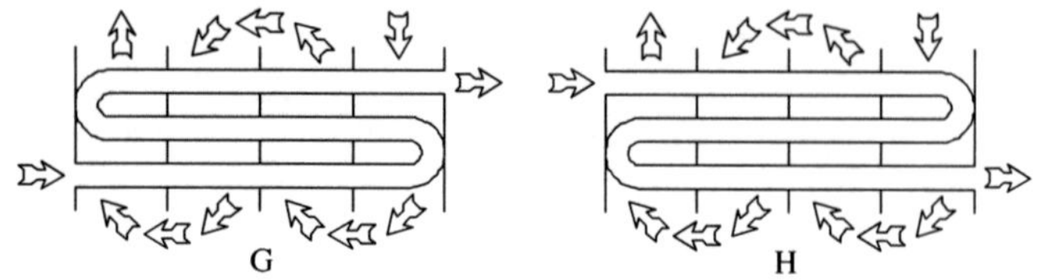

Fig. A.15

Table A.15 – Heat exchangers – G and H type – 4 passages of external fluid Corrective factor χ_c

β	ψ 0.08	0.12	0.16	0.20	0.24	0.28	0.32	0.36	0.40	0.44
0.10	*0.826*	*0.898*	*0.930*	0.950	0.962	0.971	0.977	0.982	0.986	0.988
0.15	**0.665**	*0.825*	*0.886*	0.918	0.940	0.954	0.964	0.972	0.978	0.983
0.20	....	*0.725*	*0.830*	*0.882*	0.913	0.934	0.950	0.961	0.969	0.976
0.25	....	**0.553**	*0.757*	*0.837*	0.883	0.913	0.934	0.948	0.960	0.969
0.30	....	....	*0.654*	*0.782*	0.847	0.887	0.915	0.935	0.950	0.961
0.35	....	....	**0.471**	*0.710*	*0.804*	0.858	0.894	0.920	0.938	0.952
0.40	....	....	....	*0.606*	*0.750*	0.823	0.870	0.902	0.926	0.943
0.50	....	....	....	....	*0.578*	*0.731*	0.811	0.861	0.896	0.922
0.60	....	....	....	....	....	*0.571*	*0.725*	0.807	0.859	0.896
0.70	....	....	....	....	....	....	*0.581*	0.729	0.811	0.863
0.80	....	....	....	....	....	....	**0.340**	*0.606*	0.743	0.822
0.90	....	....	....	....	....	....	....	**0.392**	*0.642*	0.765
1.00	....	....	....	....	....	....	....	....	*0.469*	0.685
1.10	....	....	....	....	....	....	....	....	**0.309**	*0.556*
1.20	....	....	....	....	....	....	....	....	....	*0.378*
1.30	....	....	....	....	....	....	....	....	....	**0.287**

If $\beta>1.3$ or $\psi<0.08$ no solution is possible in this field
Values in italics : $2<\gamma<4$; **Values in bold type : $4<\gamma<6$**

Table A.15 – (continued)

β	ψ 0.48	0.52	0.56	0.60	0.64	0.68	0.72	0.76	0.80	0.84	0.88
0.2	0.981	0.985	0.988	....	....	....	....	....	....	....	....
0.3	0.969	0.977	0.982	0.986	0.990	....	....	....	....	....	....
0.4	0.956	0.967	0.975	0.981	0.986	0.989	....	....	....	....	....
0.5	0.941	0.955	0.966	0.975	0.981	0.986	....	....	....	....	....
0.6	0.922	0.942	0.956	0.968	0.976	0.983	0.988	....	....	....	....
0.7	0.900	0.926	0.946	0.960	0.971	0.979	0.986	....	....	....	....
0.8	0.872	0.908	0.933	0.951	0.965	0.975	0.983	0.988	....	....	....
0.9	0.837	0.885	0.918	0.941	0.958	0.971	0.980	0.987	....	....	....
1.0	0.792	0.857	0.900	0.930	0.951	0.966	0.977	0.985	....	....	....
1.1	0.730	0.822	0.879	0.917	0.942	0.960	0.974	0.983	0.989	....	....
1.2	0.639	0.776	0.853	0.901	0.933	0.954	0.970	0.981	0.988	....	....
1.3	*0.494*	0.713	0.820	0.882	0.922	0.948	0.966	0.978	0.987	....	....
1.4	**0.349**	0.619	0.777	0.859	0.909	0.940	0.961	0.976	0.985	....	....
1.5	**0.283**	*0.476*	0.718	0.831	0.894	0.932	0.957	0.973	0.984	....	....
1.6	....	*0.350*	0.633	0.795	0.875	0.922	0.951	0.970	0.982	....	....
1.7	....	**0.290**	*0.505*	0.746	0.854	0.911	0.945	0.967	0.980	0.989	....
1.8	....	....	*0.378*	0.679	0.826	0.898	0.939	0.963	0.979	0.988	....
1.9	....	....	**0.310**	0.579	0.791	0.882	0.931	0.959	0.977	0.987	....
2.0	....	....	**0.272**	*0.448*	0.745	0.864	0.922	0.955	0.975	0.986	....
2.1	....	....	....	*0.354*	0.680	0.841	0.913	0.951	0.972	0.985	....
2.2	....	....	....	**0.302**	0.585	0.813	0.901	0.946	0.970	0.985	....
2.3	....	....	....	**0.274**	*0.463*	0.777	0.888	0.940	0.968	0.983	....
2.4	....	....	....	....	*0.370*	0.727	0.873	0.934	0.965	0.982	....
2.5	....	....	....	....	*0.315*	0.658	0.854	0.927	0.962	0.981	....
2.6	....	....	....	....	**0.294**	0.559	0.830	0.919	0.959	0.979	....
2.7	....	....	....	....	....	*0.446*	0.801	0.910	0.955	0.978	....
2.8	....	....	....	....	....	*0.366*	0.763	0.899	0.951	0.976	0.990
2.9	....	....	....	....	....	*0.316*	0.711	0.887	0.947	0.975	0.989
3.0	....	....	....	....	....	....	0.638	0.872	0.943	0.973	0.988

If β<0.2 or ψ> 0.88 use the mean logarithmic temperature difference for counter flow
Values in italics : 2<γ<4; **Values in bold type : 4<.γ<6**

Fig. A.16

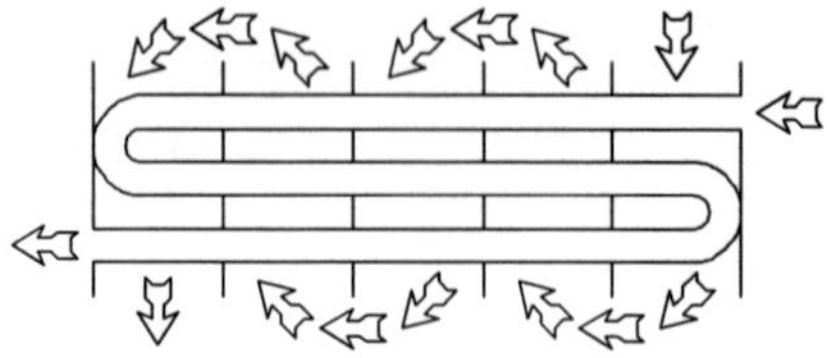

E

Table A.16 – Heat exchangers – E type – 5 passages of external fluid – Corrective factor χ_c

β \ ψ	0.04	0.08	0.12	0.16	0.20	0.24	0.28	0.32	0.36	0.40	0.44
0.05	**0.728**	*0.898*	*0.939*	0.958	0.969	0.977	0.982	0.986	0.989	....	....
0.10	....	*0.731*	*0.859*	*0.907*	0.934	0.950	0.962	0.970	0.977	0.982	0.986
0.15	....	....	*0.740*	*0.842*	0.891	0.920	0.940	0.953	0.964	0.971	0.977
0.20	....	....	....	*0.753*	*0.838*	0.884	0.914	0.934	0.949	0.960	0.969
0.25	....	....	....	*0.592*	*0.769*	0.840	0.883	0.912	0.933	0.948	0.959
0.30	....	....	....	....	*0.666*	*0.786*	0.848	0.887	0.914	0.934	0.949
0.35	....	....	....	....	....	*0.712*	0.804	0.858	0.894	0.919	0.938
0.40	....	....	....	....	....	*0.590*	*0.749*	0.823	0.870	0.902	0.926
0.45	....	....	....	....	....	....	*0.672*	0.781	0.842	0.883	0.912
0.50	....	....	....	....	....	....	*0.524*	0.727	0.810	0.861	0.897
0.60	....	....	....	....	....	....	....	*0.484*	0.720	0.807	0.860
0.70	....	....	....	....	....	....	....	....	*0.514*	0.726	0.812
0.80	....	....	....	....	....	....	....	....	....	*0.571*	0.745
0.90	....	....	....	....	....	....	....	....	....	....	0.631

If $\beta > 0.9$ no solution is possible in this field
Values in italics : $2<\gamma<4$; **Value in bold type : $4<\gamma<6$**

Table A.16 – (continued)

β	ψ										
	0.48	0.52	0.56	0.60	0.64	0.68	0.72	0.76	0.80	0.84	0.88
0.1	0.989	....	....	....	....	....	....	....	....	....	....
0.2	0.976	0.981	0.985	0.989	....	....	....	....	....	....	....
0.3	0.961	0.970	0.977	0.982	0.987	....	....	....	....	....	....
0.4	0.943	0.957	0.967	0.975	0.982	0.987	....	....	....	....	....
0.5	0.922	0.942	0.956	0.967	0.976	0.983	0.988	....	....	....	....
0.6	0.897	0.924	0.943	0.958	0.970	0.978	0.985	0.990	....	....	....
0.7	0.866	0.903	0.929	0.948	0.963	0.973	0.982	0.988	....	....	....
0.8	0.826	0.877	0.912	0.937	0.955	0.968	0.978	0.986	....	....	....
0.9	0.771	0.844	0.891	0.923	0.946	0.962	0.974	0.983	0.989	....	....
1.0	0.688	0.802	0.866	0.908	0.936	0.956	0.970	0.981	0.988	....	....
1.1	*0.463*	0.743	0.835	0.889	0.925	0.949	0.966	0.978	0.986	....	....
1.2	....	0.649	0.794	0.867	0.911	0.941	0.961	0.975	0.985	....	....
1.3	....	....	0.737	0.839	0.896	0.932	0.956	0.972	0.983	....	....
1.4	....	....	0.644	0.803	0.877	0.922	0.950	0.969	0.981	0.989	....
1.5	....	....	....	0.754	0.855	0.910	0.943	0.965	0.979	0.988	....
1.6	....	....	....	0.681	0.827	0.896	0.936	0.961	0.977	0.987	....
1.7	....	....	....	0.499	0.792	0.880	0.928	0.957	0.975	0.986	....
1.8	....	....	....	....	0.742	0.861	0.919	0.952	0.972	0.985	....
1.9	....	....	....	....	0.665	0.837	0.908	0.947	0.970	0.984	....
2.0	....	....	....	....	....	0.807	0.896	0.941	0.967	0.983	....
2.1	....	....	....	....	....	0.767	0.882	0.935	0.964	0.981	....
2.2	....	....	....	....	....	0.710	0.865	0.928	0.961	0.980	....
2.3	....	....	....	....	....	0.608	0.844	0.920	0.958	0.978	0.990
2.4	....	....	....	....	....	....	0.819	0.911	0.954	0.977	0.989
2.5	....	....	....	....	....	....	0.786	0.901	0.950	0.975	0.989
2.6	....	....	....	....	....	....	0.741	0.890	0.946	0.974	0.988
2.7	....	....	....	....	....	....	0.671	0.876	0.941	0.972	0.987
2.8	....	....	....	....	....	....	....	0.860	0.936	0.970	0.987
2.9	....	....	....	....	....	....	....	0.840	0.930	0.968	0.986
3.0	....	....	....	....	....	....	....	0.816	0.924	0.965	0.985

If $\psi > 0.88$ use the mean logarithmic temperature difference for counter flow
Value in italics : $2 < \gamma < 4$

Fig. A.17

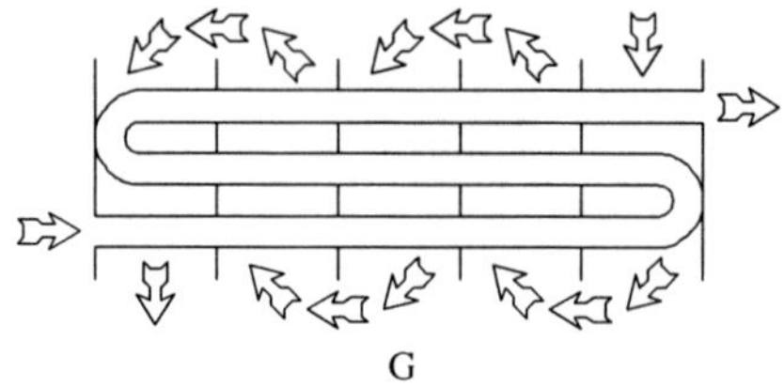

Table A.17 – Heat exchangers – G type – 5 passages of external fluid Corrective factor χ_c

β \ ψ	0.04	0.08	0.12	0.16	0.20	0.24	0.28	0.32	0.36	0.40	0.44
0.10	**0.606**	*0.844*	*0.905*	*0.935*	0.952	0.964	0.972	0.978	0.983	0.987	0.989
0.15	....	*0.718*	*0.841*	*0.893*	0.923	0.943	0.956	0.966	0.973	0.979	0.983
0.20	....	**0.522**	*0.756*	*0.843*	*0.889*	0.919	0.938	0.952	0.963	0.971	0.977
0.25	....	....	*0.638*	*0.781*	*0.849*	0.890	0.918	0.937	0.951	0.962	0.970
0.30	....	....	**0.471**	*0.698*	*0.800*	0.857	0.894	0.919	0.938	0.952	0.963
0.40	....	....	....	**0.441**	*0.659*	*0.771*	0.836	0.878	0.908	0.930	0.946
0.50	....	....	....	....	**0.426**	*0.636*	*0.754*	0.823	0.870	0.902	0.926
0.60	....	....	....	....	....	**0.424**	*0.628*	*0.747*	0.820	0.867	0.901
0.70	....	....	....	....	....	....	**0.434**	*0.632*	0.751	0.823	0.871
0.80	....	....	....	....	....	....	....	*0.458*	*0.649*	0.762	0.833
0.90	....	....	....	....	....	....	....	**0.330**	*0.496*	*0.676*	0.781
1.00	....	....	....	....	....	....	....	....	**0.356**	*0.547*	0.711
1.10	....	....	....	....	....	....	....	....	....	**0.398**	*0.608*
1.20	....	....	....	....	....	....	....	....	....	**0.319**	*0.466*
1.30	....	....	....	....	....	....	....	....	....	....	**0.357**
1.40	....	....	....	....	....	....	....	....	....	....	**0.304**

If β> 1.4 no solution is possible in this field
Values in italics : 2<γ<4 ; **Values in bold type : 4<γ< 6**

Table A.17 – (continued)

	ψ										
β	0.48	0.52	0.56	0.60	0.64	0.68	0.72	0.76	0.80	0.84	0.88
0.2	0.982	0.986	0.989	....	....	....	....	....	....	....	....
0.3	0.971	0.977	0.983	0.987	....	....	....	....	....	....	....
0.4	0.958	0.968	0.976	0.982	0.986	....	....	....	....	....	....
0.5	0.944	0.957	0.968	0.976	0.982	0.987	....	....	....	....	....
0.6	0.926	0.944	0.958	0.969	0.977	0.984	0.989	....	....	....	....
0.7	0.905	0.930	0.948	0.962	0.972	0.980	0.986	....	....	....	....
0.8	0.879	0.912	0.936	0.953	0.966	0.976	0.984	0.989	....	....	....
0.9	0.847	0.891	0.922	0.944	0.960	0.972	0.981	0.987	....	....	....
1.0	0.806	0.865	0.905	0.933	0.953	0.967	0.978	0.985	....	....	....
1.1	0.750	0.833	0.885	0.921	0.945	0.963	0.975	0.984	0.990	....	....
1.2	0.672	0.791	0.861	0.906	0.936	0.957	0.971	0.981	0.989	....	....
1.3	*0.559*	0.735	0.831	0.888	0.926	0.950	0.967	0.979	0.987	....	....
1.4	*0.426*	0.655	0.792	0.867	0.913	0.943	0.963	0.977	0.986	....	....
1.5	**0.343**	*0.541*	0.739	0.841	0.899	0.935	0.959	0.974	0.984	....	....
1.6	**0.299**	*0.420*	0.666	0.808	0.882	0.926	0.954	0.971	0.983	....	....
1.7	....	**0.345**	*0.562*	0.764	0.862	0.915	0.948	0.968	0.981	0.990	....
1.8	....	**0.303**	*0.444*	0.705	0.836	0.903	0.941	0.965	0.980	0.989	....
1.9	....	....	*0.364*	0.621	0.804	0.889	0.934	0.961	0.978	0.988	....
2.0	....	....	**0.317**	*0.511*	0.762	0.871	0.926	0.957	0.976	0.987	....
2.1	....	....	....	*0.410*	0.706	0.850	0.917	0.953	0.974	0.986	....
2.2	....	....	....	*0.348*	0.626	0.824	0.906	0.948	0.971	0.985	....
2.3	....	....	....	**0.309**	0.521	0.791	0.894	0.943	0.969	0.984	....
2.4	....	....	....	....	*0.423*	0.747	0.879	0.937	0.967	0.983	....
2.5	....	....	....	....	*0.360*	0.687	0.862	0.930	0.963	0.982	....
2.6	....	....	....	....	**0.318**	0.603	0.840	0.923	0.961	0.980	....
2.7	....	....	....	....	....	0.500	0.814	0.914	0.958	0.979	....
2.8	....	....	....	....	....	*0.413*	0.779	0.904	0.954	0.978	....
2.9	....	....	....	....	....	*0.356*	0.733	0.893	0.950	0.976	0.989
3.0	....	....	....	....	....	**0.322**	0.670	0.879	0.945	0.975	0.989

If $\beta < 0.2$ or $\psi > 0.88$ use the mean logarithmic temperature difference for counter flow
Values in italics : $2 < \gamma < 4$; **Values in bold type :** $4 < \gamma < 6$

A.2.3 Heat Exchangers with 4 Passages of Internal Fluid (Fig. 2.7)

Fig. A.18

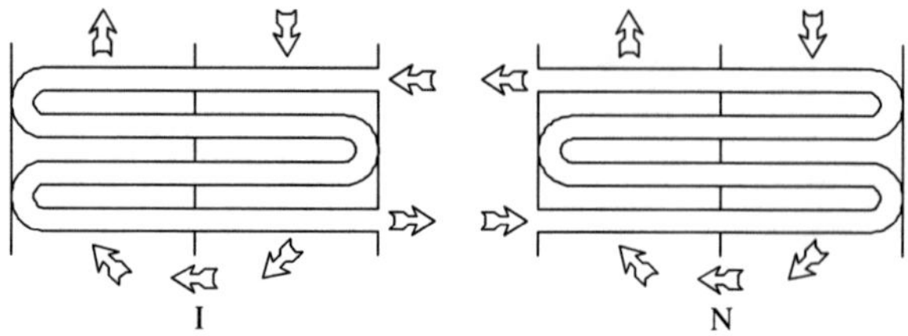

Table A.18 – Heat exchangers – I and N type – 2 passages of external fluid Corrective factor χ_c

β \ ψ	0.04	0.08	0.12	0.16	0.20	0.24	0.28	0.32	0.36	0.40	0.44
0.05	**0.761**	*0.907*	*0.944*	0.962	0.972	0.979	0.984	0.987	0.990	....	....
0.10	....	*0.764*	*0.872*	*0.915*	0.939	0.955	0.965	0.973	0.979	0.983	0.987
0.15	....	....	*0.771*	*0.858*	0.901	0.927	0.945	0.957	0.967	0.974	0.979
0.20	....	....	**0.560**	*0.782*	*0.854*	0.895	0.921	0.940	0.954	0.964	0.972
0.25	....	....	....	*0.666*	*0.795*	0.857	0.894	0.920	0.939	0.953	0.963
0.30	....	....	....	....	*0.714*	*0.810*	0.863	0.898	0.922	0.940	0.954
0.35	....	....	....	....	*0.573*	*0.750*	0.825	0.872	0.904	0.927	0.943
0.40	....	....	....	....	....	*0.665*	0.779	0.842	0.883	0.911	0.932
0.45	....	....	....	....	....	**0.450**	*0.720*	0.806	0.859	0.894	0.920
0.50	....	....	....	....	....	....	*0.634*	0.762	0.830	0.875	0.907
0.60	....	....	....	....	....	....	....	*0.623*	0.757	0.828	0.874
0.70	....	....	....	....	....	....	....	....	*0.633*	0.762	0.833
0.80	....	....	....	....	....	....	....	....	....	0.658	0.777
0.90	....	....	....	....	....	....	....	....	....	....	0.694
1.00	....	....	....	....	....	....	....	....	....	....	*0.516*

If β> 1.0 no solution is possible in this field
Values in italics : 2<γ<4 ; **Value in bold type : 4<γ<6**

Table A.18 – (continued)

β	ψ 0.48	0.52	0.56	0.60	0.64	0.68	0.72	0.76	0.80	0.84	0.88
0.2	0.978	0.983	0.986	0.990	....	....	....	....	....	....	....
0.3	0.964	0.972	0.979	0.984	0.988	....	....	....	....	....	....
0.4	0.948	0.960	0.970	0.977	0.983	0.988	....	....	....	....	....
0.5	0.930	0.947	0.960	0.970	0.978	0.984	0.989	....	....	....	....
0.6	0.907	0.931	0.949	0.962	0.972	0.980	0.988	....	....	....	....
0.7	0.879	0.912	0.935	0.953	0.966	0.976	0.983	0.989	....	....	....
0.8	0.845	0.889	0.920	0.942	0.959	0.971	0.980	0.987	....	....	....
0.9	0.799	0.861	0.902	0.930	0.951	0.966	0.977	0.985	....	....	....
1.0	0.735	0.825	0.880	0.916	0.942	0.960	0.973	0.982	0.989	....	....
1.1	0.630	0.777	0.853	0.900	0.932	0.954	0.969	0.980	0.987	....	....
1.2	....	0.708	0.818	0.880	0.920	0.946	0.964	0.977	0.986	....	....
1.3	....	0.584	0.773	0.856	0.906	0.938	0.960	0.974	0.984	....	....
1.4	....	....	0.707	0.826	0.890	0.929	0.954	0.971	0.983	....	....
1.5	....	....	0.589	0.787	0.871	0.919	0.949	0.968	0.981	0.989	....
1.6	....	....	....	0.733	0.847	0.906	0.942	0.964	0.979	0.988	....
1.7	....	....	....	0.646	0.817	0.892	0.935	0.961	0.977	0.987	....
1.8	....	....	....	....	0.778	0.876	0.927	0.957	0.975	0.986	....
1.9	....	....	....	....	0.723	0.856	0.917	0.952	0.973	0.985	....
2.0	....	....	....	....	0.633	0.830	0.907	0.947	0.970	0.984	....
2.1	....	....	....	....	....	0.799	0.894	0.941	0.968	0.983	....
2.2	....	....	....	....	....	0.755	0.880	0.935	0.965	0.982	....
2.3	....	....	....	....	....	0.692	0.862	0.928	0.962	0.980	....
2.4	....	....	....	....	....	0.568	0.841	0.920	0.958	0.979	....
2.5	....	....	....	....	....	....	0.814	0.911	0.955	0.977	0.990
2.6	....	....	....	....	....	....	0.779	0.901	0.951	0.976	0.989
2.7	....	....	....	....	....	....	0.731	0.889	0.947	0.974	0.988
2.8	....	....	....	....	....	....	0.653	0.875	0.942	0.972	0.988
2.9	....	....	....	....	....	....	....	0.859	0.937	0.971	0.987
3.0	....	....	....	....	....	....	....	0.839	0.931	0.969	0.986

If $\beta < 0.2$ or $\psi > 0.88$ use the mean logarithmic temperature difference for counter flow

Fig. A.19

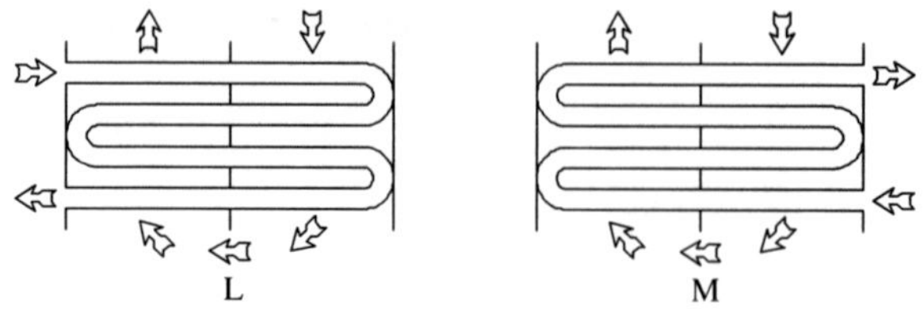

Table A.19 – Heat exchangers – L and M type – 2 passages of external fluid – Corrective factor χ_c

β \ ψ	0.08	0.12	0.16	0.20	0.24	0.28	0.32	0.36	0.40	0.44
0.10	*0.823*	*0.894*	*0.927*	0.947	0.960	0.969	0.976	0.981	0.985	0.988
0.15	**0.676**	*0.820*	*0.881*	0.914	0.936	0.951	0.962	0.970	0.976	0.982
0.20	....	*0.722*	*0.823*	*0.876*	0.909	0.931	0.946	0.958	0.967	0.974
0.25	....	**0.582**	*0.751*	*0.830*	0.877	0.908	0.929	0.945	0.957	0.967
0.30	....	....	*0.654*	*0.774*	0.839	0.881	0.910	0.931	0.946	0.958
0.35	....	....	**0.520**	*0.702*	*0.795*	0.851	0.888	0.915	0.934	0.949
0.40	....	....	....	*0.607*	*0.740*	0.815	0.863	0.897	0.921	0.940
0.45	....	....	....	**0.481**	*0.670*	*0.772*	0.834	0.877	0.907	0.929
0.50	....	....	....	....	*0.578*	*0.719*	0.801	0.853	0.890	0.917
0.60	....	....	....	....	**0.351**	*0.566*	*0.711*	0.796	0.851	0.889
0.70	....	....	....	....	....	**0.353**	*0.570*	0.715	0.799	0.855
0.80	....	....	....	....	....	....	**0.367**	*0.589*	0.728	0.810
0.90	....	....	....	....	....	....	....	**0.397**	*0.621*	0.750
1.00	....	....	....	....	....	....	....	....	*0.450*	0.664
1.10	....	....	....	....	....	....	....	....	**0.298**	*0.526*
1.20	....	....	....	....	....	....	....	....	....	**0.350**

If $\beta > 1.2$ or $\psi < 0.08$ no solution is possible in this field
Values in italics : $2<\gamma<4$; **Values in bold type : $4<\gamma<6$**

Table A.19 – (continued)

β \ ψ	0.48	0.52	0.56	0.60	0.64	0.68	0.72	0.76	0.80	0.84	0.88
0.2	0.980	0.984	0.988	....	....	....	....	....	....	....	....
0.3	0.968	0.975	0.981	0.985	0.989	....	....	....	....	....	....
0.4	0.954	0.964	0.973	0.980	0.985	0.989	....	....	....	....	....
0.5	0.937	0.952	0.964	0.973	0.980	0.986	0.990	....	....	....	....
0.6	0.917	0.938	0.954	0.966	0.975	0.982	0.987	....	....	....	....
0.7	0.894	0.921	0.942	0.958	0.969	0.978	0.985	0.990	....	....	....
0.8	0.864	0.902	0.928	0.948	0.963	0.974	0.982	0.988	....	....	....
0.9	0.827	0.878	0.912	0.937	0.956	0.969	0.979	0.986	....	....	....
1.0	0.778	0.848	0.894	0.925	0.948	0.964	0.976	0.984	0.990	....	....
1.1	0.711	0.810	0.871	0.911	0.939	0.958	0.972	0.982	0.989	....	....
1.2	0.611	0.760	0.843	0.894	0.928	0.952	0.968	0.979	0.987	....	....
1.3	*0.451*	0.690	0.807	0.874	0.916	0.944	0.964	0.977	0.986	....	....
1.4	**0.306**	0.583	0.760	0.849	0.902	0.936	0.959	0.974	0.984	....	....
1.5	....	*0.419*	0.693	0.818	0.886	0.927	0.954	0.971	0.983	....	....
1.6	....	**0.293**	0.594	0.778	0.866	0.917	0.948	0.968	0.981	0.989	....
1.7	....	....	*0.438*	0.724	0.842	0.905	0.942	0.964	0.979	0.988	....
1.8	....	....	**0.306**	0.644	0.812	0.890	0.934	0.961	0.977	0.987	....
1.9	....	....	**0.240**	0.519	0.773	0.874	0.926	0.957	0.975	0.986	....
2.0	....	....	....	*0.362*	0.719	0.853	0.917	0.952	0.973	0.986	....
2.1	....	....	....	**0.272**	0.641	0.828	0.906	0.947	0.971	0.985	....
2.2	....	....	....	....	0.517	0.797	0.894	0.942	0.968	0.983	....
2.3	....	....	....	....	*0.365*	0.755	0.880	0.936	0.965	0.982	....
2.4	....	....	....	....	**0.275**	0.696	0.862	0.929	0.963	0.981	....
2.5	....	....	....	....	**0.223**	0.607	0.841	0.921	0.959	0.980	....
2.6	....	....	....	....	....	0.468	0.815	0.913	0.956	0.978	....
2.7	....	....	....	....	....	*0.334*	0.782	0.903	0.952	0.977	0.989
2.8	....	....	....	....	....	*0.261*	0.736	0.891	0.948	0.975	0.989
2.9	....	....	....	....	....	**0.213**	0.671	0.878	0.944	0.973	0.988
3.0	....	....	....	....	....	....	0.570	0.861	0.939	0.972	0.988

If $\beta < 0.2$ or $\psi > 0.88$ use the mean logarithmic temperature difference for counter flow
Values in italics : $2<\gamma<4$; **Values in bold type : $4<\gamma<6$**

Fig. A.20

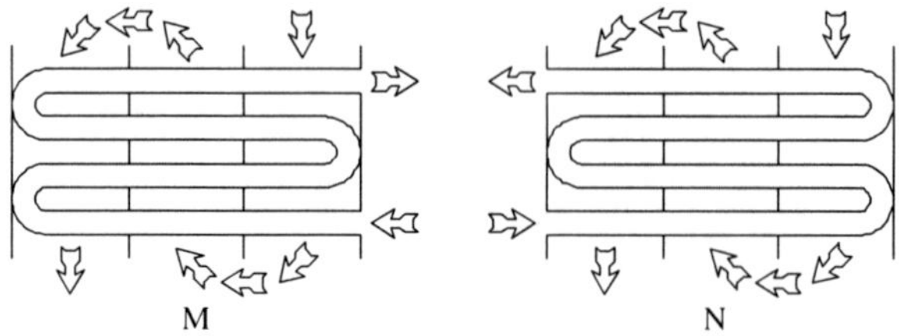

Table A.20 – Heat exchangers – M and N type – 3 passages of external fluid Corrective factor χ_c

	ψ										
β	0.04	0.08	0.12	0.16	0.20	0.24	0.28	0.32	0.36	0.40	0.44
0.10	**0.675**	*0.848*	*0.905*	*0.934*	0.951	0.963	0.971	0.978	0.982	0.986	0.989
0.15	....	*0.735*	*0.841*	*0.892*	0.921	0.941	0.955	0.965	0.972	0.978	0.983
0.20	....	**0.591**	*0.761*	*0.841*	*0.887*	0.916	0.936	0.950	0.961	0.969	0.976
0.25	....	....	*0.657*	*0.780*	*0.846*	0.887	0.915	0.934	0.949	0.960	0.969
0.30	....	....	**0.532**	*0.701*	*0.797*	0.853	0.890	0.916	0.936	0.950	0.961
0.35	....	....	....	*0.601*	*0.735*	*0.813*	0.863	0.896	0.921	0.939	0.953
0.40	....	....	....	**0.484**	*0.658*	*0.765*	0.830	0.873	0.904	0.927	0.943
0.45	....	....	....	....	*0.558*	*0.705*	*0.791*	0.847	0.885	0.913	0.934
0.50	....	....	....	....	**0.446**	*0.627*	*0.744*	0.816	0.864	0.898	0.922
0.60	....	....	....	....	....	**0.418**	*0.610*	*0.735*	0.811	0.861	0.897
0.70	....	....	....	....	....	....	**0.402**	*0.608*	0.736	0.813	0.864
0.80	....	....	....	....	....	....	....	**0.400**	*0.620*	0.747	0.823
0.90	....	....	....	....	....	....	....	....	*0.419*	*0.647*	0.767
1.00	....	....	....	....	....	....	....	....	....	*0.468*	0.684
1.10	....	....	....	....	....	....	....	....	....	**0.282**	*0.545*
1.20	....	....	....	....	....	....	....	....	....	....	**0.323**

If β>1.2 no solution is possible in this field
Values in italics : 2<γ<4 ; **Values in bold type : 4<γ<6**

Table A.20 – (continued)

β	ψ 0.48	0.52	0.56	0.60	0.64	0.68	0.72	0.76	0.80	0.84	0.88
0.2	0.981	0.985	0.989	....	....	....	....	....	....	....	....
0.3	0.970	0.977	0.982	0.986	0.990	....	....	....	....	....	....
0.4	0.957	0.967	0.975	0.981	0.986	0.989	....	....	....	....	....
0.5	0.941	0.955	0.966	0.975	0.981	0.986	....	....	....	....	....
0.6	0.923	0.942	0.957	0.968	0.976	0.983	0.988	....	....	....	....
0.7	0.900	0.926	0.946	0.960	0.971	0.979	0.986	....	....	....	....
0.8	0.873	0.908	0.933	0.951	0.965	0.975	0.983	0.989	....	....	....
0.9	0.838	0.885	0.918	0.941	0.958	0.971	0.980	0.987	....	....	....
1.0	0.792	0.857	0.900	0.930	0.951	0.966	0.977	0.985	....	....	....
1.1	0.728	0.822	0.879	0.916	0.942	0.961	0.974	0.983	0.989	....	....
1.2	0.629	0.774	0.852	0.900	0.933	0.954	0.970	0.981	0.988	....	....
1.3	*0.440*	0.706	0.818	0.881	0.921	0.948	0.966	0.978	0.987	....	....
1.4	**0.260**	0.595	0.772	0.858	0.908	0.940	0.961	0.976	0.985	....	....
1.5	....	*0.376*	0.708	0.828	0.893	0.931	0.956	0.973	0.984	....	....
1.6	....	**0.237**	0.603	0.790	0.874	0.922	0.951	0.970	0.982	....	....
1.7	....	....	*0.386*	0.736	0.851	0.910	0.945	0.967	0.980	0.989	....
1.8	....	....	**0.237**	0.655	0.822	0.897	0.938	0.963	0.979	0.988	....
1.9	....	....	....	*0.501*	0.784	0.881	0.931	0.959	0.977	0.987	....
2.0	....	....	....	**0.274**	0.731	0.861	0.922	0.955	0.975	0.986	....
2.1	....	....	....	**0.203**	0.649	0.837	0.912	0.950	0.972	0.985	....
2.2	....	....	....	....	0.487	0.806	0.900	0.945	0.970	0.984	....
2.3	....	....	....	....	*0.266*	0.765	0.886	0.939	0.967	0.983	....
2.4	....	....	....	....	**0.199**	0.705	0.870	0.933	0.965	0.982	....
2.5	....	....	....	....	....	0.606	0.849	0.926	0.962	0.981	....
2.6	....	....	....	....	....	**0.381**	0.824	0.917	0.958	0.979	....
2.7	....	....	....	....	....	**0.235**	0.791	0.908	0.955	0.978	....
2.8	....	....	....	....	....	**0.185**	0.745	0.897	0.951	0.976	0.989
2.9	....	....	....	....	....	....	0.676	0.884	0.947	0.975	0.989
3.0	....	....	....	....	....	....	0.548	0.868	0.942	0.973	0.988

If $\beta < 0.2$ or $\psi > 0.88$ use the mean logarithmic temperature difference for counter flow
Values in italics : $2<\gamma<4$; **Values in bold type : $4<\gamma<6$**

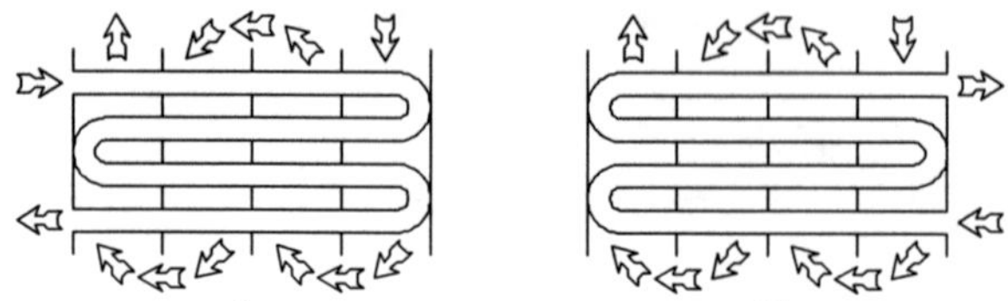

L M

Fig. A.21

Table A.21 – Heat exchangers – L and M type – 4 passages of external fluid – Corrective factor χ_c

β \ ψ	0.04	0.08	0.12	0.16	0.20	0.24	0.28	0.32	0.36	0.40	0.44
0.05	**0.813**	*0.918*	*0.949*	0.965	0.974	0.980	0.985	0.988	....	....	....
0.10	....	*0.801*	*0.885*	*0.923*	0.944	0.958	0.968	0.975	0.980	0.984	0.988
0.15	....	**0.584**	*0.801*	*0.871*	0.909	0.933	0.949	0.960	0.969	0.976	0.981
0.20	....	....	*0.674*	*0.806*	*0.867*	0.903	0.927	0.944	0.956	0.966	0.973
0.25	....	....	....	*0.717*	*0.815*	0.868	0.902	0.926	0.943	0.955	0.965
0.30	....	....	....	*0.570*	*0.748*	*0.827*	0.874	0.905	0.927	0.944	0.957
0.35	....	....	....	....	*0.654*	*0.775*	0.840	0.881	0.910	0.931	0.947
0.40	....	....	....	....	**0.474**	*0.707*	0.799	0.854	0.891	0.917	0.937
0.45	....	....	....	....	....	*0.608*	*0.748*	0.822	0.869	0.901	0.925
0.50	....	....	....	....	....	**0.388**	*0.681*	0.783	0.843	0.884	0.913
0.60	....	....	....	....	....	....	**0.346**	*0.671*	0.778	0.841	0.883
0.70	....	....	....	....	....	....	....	**0.365**	*0.676*	0.782	0.845
0.80	....	....	....	....	....	....	....	....	*0.441*	0.695	0.795
0.90	....	....	....	....	....	....	....	....	....	*0.527*	0.723
1.00	....	....	....	....	....	....	....	....	....	....	*0.603*
1.10	....	....	....	....	....	....	....	....	....	....	**0.255**

If β> 1.1 no solution is possible in this field

Values in italics : 2<γ<4 ; **Values in bold type : 4<γ<6**

Table A.21 – (continued)

β \ ψ	0.48	0.52	0.56	0.60	0.64	0.68	0.72	0.76	0.80	0.84	0.88
0.2	0.979	0.984	0.987	....	....	....	....	....	....	....	....
0.3	0.966	0.974	0.980	0.985	0.989	....	....	....	....	....	....
0.4	0.951	0.963	0.972	0.979	0.984	0.988	....	....	....	....	....
0.5	0.934	0.950	0.962	0.972	0.979	0.985	0.989	....	....	....	....
0.6	0.913	0.935	0.952	0.964	0.974	0.981	0.987	....	....	....	....
0.7	0.887	0.917	0.939	0.956	0.968	0.977	0.984	0.989	....	....	....
0.8	0.855	0.896	0.925	0.946	0.961	0.973	0.981	0.987	....	....	....
0.9	0.814	0.870	0.908	0.934	0.954	0.968	0.978	0.985	....	....	....
1.0	0.758	0.837	0.888	0.922	0.945	0.962	0.974	0.983	0.989	....	....
1.1	0.673	0.795	0.863	0.906	0.936	0.956	0.971	0.981	0.988	....	....
1.2	*0.501*	0.735	0.832	0.888	0.925	0.950	0.967	0.978	0.987	....	....
1.3	....	0.642	0.791	0.866	0.912	0.942	0.962	0.976	0.985	....	....
1.4	....	*0.417*	0.734	0.839	0.897	0.933	0.957	0.973	0.984	....	....
1.5	....	....	0.647	0.804	0.879	0.924	0.952	0.970	0.982	....	....
1.6	....	....	*0.450*	0.757	0.858	0.912	0.946	0.967	0.980	0.989	....
1.7	....	....	**0.201**	0.688	0.831	0.899	0.939	0.963	0.978	0.988	....
1.8	....	....	....	0.566	0.796	0.884	0.931	0.959	0.976	0.987	....
1.9	....	....	....	**0.251**	0.750	0.866	0.922	0.955	0.974	0.986	....
2.0	....	....	....	....	0.681	0.843	0.913	0.950	0.972	0.985	....
2.1	....	....	....	....	0.558	0.815	0.901	0.945	0.969	0.984	....
2.2	....	....	....	....	*0.250*	0.778	0.888	0.939	0.967	0.983	....
2.3	....	....	....	....	**0.176**	0.726	0.872	0.933	0.964	0.981	....
2.4	....	....	....	....	....	0.647	0.853	0.925	0.961	0.980	....
2.5	....	....	....	....	....	0.474	0.829	0.917	0.958	0.979	....
2.6	....	....	....	....	....	**0.218**	0.799	0.908	0.954	0.977	0.990
2.7	....	....	....	....	....	**0.165**	0.759	0.897	0.950	0.976	0.989
2.8	....	....	....	....	....	....	0.701	0.884	0.946	0.974	0.988
2.9	....	....	....	....	....	....	0.605	0.870	0.941	0.972	0.988
3.0	....	....	....	....	....	....	*0.338*	0.852	0.936	0.970	0.987

If $\beta < 0.2$ or $\psi > 0.88$ use the mean logarithmic temperature difference for counter flow
Values in italics : $2 < \gamma < 4$; **Values in bold type :** $4 < \gamma < 6$

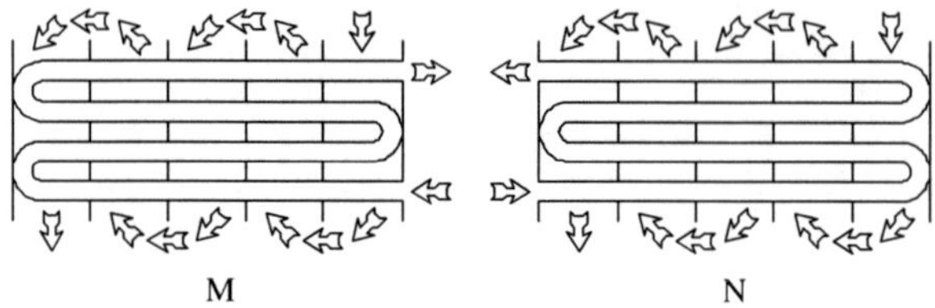

Fig. A.22

Table A.22 – Heat exchangers – M and N type – 5 passages of external fluid – Corrective factor χ_c

β \ ψ	0.04	0.08	0.12	0.16	0.20	0.24	0.28	0.32	0.36	0.40	0.44
0.05	**0.831**	*0.922*	*0.951*	0.966	0.975	0.981	0.985	0.988	....	....	....
0.10	....	*0.816*	*0.891*	*0.926*	0.946	0.959	0.969	0.976	0.981	0.985	0.988
0.15	....	**0.647**	*0.813*	*0.877*	0.912	0.935	0.950	0.962	0.970	0.976	0.981
0.20	....	....	*0.703*	*0.816*	*0.872*	0.907	0.929	0.946	0.958	0.967	0.974
0.25	....	....	**0.513**	*0.736*	*0.824*	0.874	0.906	0.928	0.945	0.957	0.966
0.30	....	....	....	*0.617*	*0.762*	*0.834*	0.878	0.908	0.930	0.946	0.958
0.35	....	....	....	**0.394**	*0.679*	*0.786*	0.846	0.885	0.913	0.934	0.949
0.40	....	....	....	....	*0.547*	*0.723*	0.807	0.859	0.894	0.920	0.939
0.45	....	....	....	....	....	*0.637*	*0.760*	0.828	0.873	0.904	0.927
0.50	....	....	....	....	....	*0.491*	*0.699*	0.792	0.849	0.887	0.915
0.60	....	....	....	....	....	....	*0.457*	*0.688*	0.786	0.846	0.887
0.70	....	....	....	....	....	....	....	*0.459*	*0.692*	0.790	0.850
0.80	....	....	....	....	....	....	....	....	*0.499*	0.708	0.802
0.90	....	....	....	....	....	....	....	....	....	*0.559*	0.734
1.00	....	....	....	....	....	....	....	....	....	....	*0.623*
1.10	....	....	....	....	....	....	....	....	....	....	**0.252**

If β> 1.1 no solution is possible in this field
Values in italics : 2<γ<4 ; **Values in bold type : 4<γ<6**

Table A.22 – (continued)

β	ψ 0.48	0.52	0.56	0.60	0.64	0.68	0.72	0.76	0.80	0.84	0.88
0.2	0.980	0.984	0.988	....	....	....	....	....	....	....	....
0.3	0.967	0.975	0.981	0.985	0.989	....	....	....	....	....	....
0.4	0.953	0.964	0.972	0.979	0.985	0.989	....	....	....	....	....
0.5	0.936	0.951	0.963	0.973	0.980	0.985	0.990	....	....	....	....
0.6	0.916	0.937	0.953	0.965	0.975	0.982	0.987	....	....	....	....
0.7	0.891	0.920	0.941	0.957	0.969	0.978	0.985	0.990	....	....	....
0.8	0.860	0.899	0.927	0.947	0.962	0.973	0.982	0.988	....	....	....
0.9	0.820	0.874	0.911	0.937	0.955	0.969	0.979	0.986	....	....	....
1.0	0.766	0.842	0.891	0.924	0.947	0.963	0.975	0.984	0.990	....	....
1.1	0.686	0.801	0.867	0.909	0.937	0.957	0.971	0.981	0.988	....	....
1.2	*0.531*	0.745	0.837	0.891	0.927	0.951	0.967	0.979	0.987	....	....
1.3	....	0.657	0.797	0.870	0.914	0.943	0.963	0.976	0.986	....	....
1.4	....	*0.454*	0.743	0.844	0.900	0.935	0.958	0.974	0.984	....	....
1.5	....	....	0.660	0.810	0.883	0.926	0.953	0.971	0.982	....	....
1.6	....	....	*0.479*	0.764	0.862	0.915	0.947	0.968	0.981	0.989	....
1.7	....	....	....	0.699	0.836	0.902	0.940	0.964	0.979	0.988	....
1.8	....	....	....	0.583	0.803	0.887	0.933	0.960	0.977	0.987	....
1.9	....	....	....	**0.210**	0.758	0.870	0.925	0.956	0.975	0.986	....
2.0	....	....	....	....	0.692	0.848	0.915	0.951	0.973	0.985	....
2.1	....	....	....	....	0.573	0.820	0.904	0.946	0.970	0.984	....
2.2	....	....	....	....	**0.205**	0.784	0.891	0.941	0.968	0.983	....
2.3	....	....	....	....	....	0.735	0.876	0.934	0.965	0.982	....
2.4	....	....	....	....	....	0.658	0.857	0.927	0.962	0.981	....
2.5	....	....	....	....	....	0.488	0.834	0.919	0.959	0.979	....
2.6	....	....	....	....	....	**0.177**	0.805	0.910	0.955	0.978	....
2.7	....	....	....	....	....	....	0.766	0.900	0.951	0.976	0.989
2.8	....	....	....	....	....	....	0.710	0.888	0.947	0.975	0.989
2.9	....	....	....	....	....	....	0.616	0.873	0.943	0.973	0.988
3.0	....	....	....	....	....	....	*0.278*	0.856	0.937	0.971	0.987

If $\beta < 0.2$ or $\psi > 0.88$ use the mean logarithmic temperature difference for counter flow
Values in italics : $2<\gamma<4$; **Values in bold type : $4<\gamma<6$**

A.3 Coils

A.3.1 Coils with Fluids in Parallel Flow (Fig. 2.8)

Fig. A.23

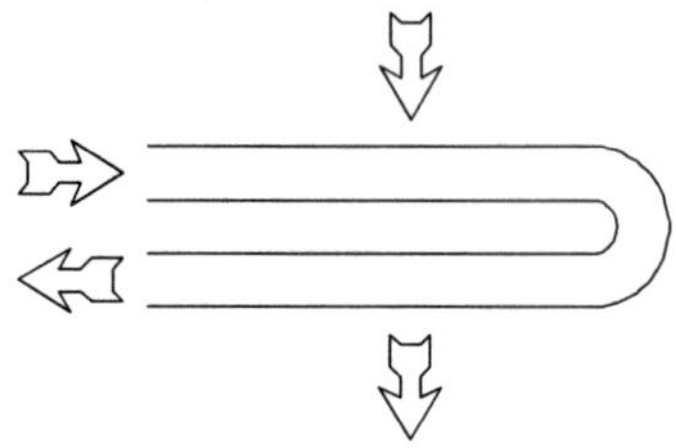

Table A.23 – Coils – parallel flow – 2 sections – Corrective factor χ_p

	ψ								
β	0.12	0.16	0.20	0.24	0.28	0.32	0.36	0.40	0.44
0.10	*1.044*	*1.024*	1.016	1.012	....	....	....	....	....
0.15	....	*1.057*	*1.032*	1.021	1.015	1.012	....	....	....
0.20	....	....	*1.061*	1.035	1.023	1.017	1.013	1.010	....
0.25	....	....	....	*1.060*	1.036	1.024	1.018	1.013	1.010
0.30	....	....	....	*1.123*	*1.056*	1.035	1.024	1.017	1.013
0.35	....	....	....	....	*1.093*	1.049	1.032	1.022	1.016
0.40	....	....	....	....	....	*1.073*	1.043	1.028	1.020
0.45	....	....	....	....	....	*1.114*	1.059	1.037	1.025
0.50	....	....	....	....	....	....	1.083	1.047	1.030
0.60	....	....	....	....	....	....	....	1.084	1.047
0.70	....	....	....	....	....	....	....	....	1.077

If $\beta > 0.7$ or $\psi < 0.12$ no solution is possible in this field
Values in italics : $2<\gamma<4$

Table A.23 – (continued)

β	ψ								
	0.48	0.52	0.56	0.60	0.64	0.68	0.72	0.76	0.80
0.3	1.010	....	....	....	....	....	....	....	....
0.4	1.015	1.011	....	....	....	....	....	....	....
0.5	1.021	1.015	1.011	....	....	....	....	....	....
0.6	1.030	1.021	1.015	1.011	....	....	....	....	....
0.7	1.044	1.028	1.019	1.013	....	....	....	....	....
0.8	1.066	1.039	1.025	1.017	1.011	....	....	....	....
0.9	....	1.053	1.033	1.021	1.014	....	....	....	....
1.0	....	1.073	1.042	1.026	.1.017	1.011	....	....	....
1.1	....	....	1.057	1.033	1.020	1.013	....	....	....
1.2	....	....	1.058	1.041	1.024	1.015	1.010	....	....
1.3	....	....	....	1.051	1.029	1.018	1.011	....	....
1.4	....	....	....	1.054	1.036	1.021	1.013	....	....
1.5	....	....	....	....	1.043	1.024	1.014	....	....
1.6	....	....	....	....	1.050	1.028	1.016	1.010	....
1.7	....	....	....	....	....	1.034	1.018	1.011	....
1.8	....	....	....	....	....	1.039	1.021	1.012	....
1.9	....	....	....	....	....	1.044	1.024	1.013	....
2.0	....	....	....	....	....	1.029	1.027	1.015	....
2.2	....	....	....	....	....	....	1.036	1.018	....
2.4	....	....	....	....	....	....	1.028	1.022	1.011
2.6	....	....	....	....	....	....	....	1.027	1.013
2.8	....	....	....	....	....	....	....	1.032	1.015
3.0	....	....	....	....	....	....	....	....	1.018

If β <0.3 or ψ >0.8 use the mean logarithmic temperature difference for parallel flow

Fig. A.24

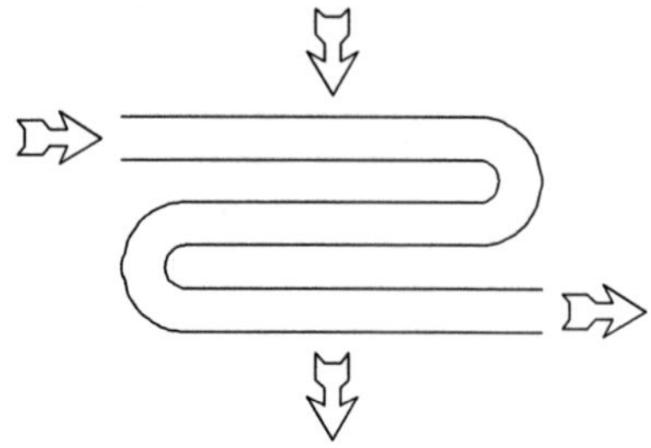

Table A.24 – Coils – parallel flow – 3 sections – Corrective factor χ_p

β \ ψ	0.12	0.16	0.20	0.24	0.28	0.32	0.36	0.40	0.44
0.10	*1.019*	*1.011*	....	....	....	....	....	....	....
0.15	....	*1.025*	*1.014*	1.010	....	....	....	....	....
0.20	....	....	*1.028*	1.016	1.011	....	....	....	....
0.25	....	....	....	*1.028*	1.016	1.011	....	....	....
0.30	....	....	....	*1.069*	*1.026*	1.016	1.011	....	....
0.35	....	....	....	....	*1.048*	1.023	1.015	1.010	....
0.40	....	....	....	....	....	*1.037*	1.020	1.013	....
0.45	....	....	....	....	....	*1.078*	1.028	1.017	1.012
0.50	....	....	....	....	....	....	*1.045*	1.023	1.014
0.55	....	....	....	....	....	....	*1.112*	1.032	1.018
0.60	....	....	....	....	....	....	....	1.048	1.023
0.65	....	....	....	....	....	....	....	*1.106*	1.030
0.70	....	....	....	....	....	....	....	....	1.044
0.75	....	....	....	....	....	....	....	....	*1.077*

If $\beta > 0.75$ or $\psi < 0.12$ no solution is possible in this field
Values in italics : $2 < \gamma < 4$

Table A.24 – (continued)

β	ψ 0.48	0.52	0.56	0.60	0.64	0.68	0.72	0.76	0.80
0.50	1.010	....	....	....	....	....	....	....	....
0.55	1.012	....	....	....	....	....	....	....	....
0.60	1.014	1.010	....	....	....	....	....	....	....
0.65	1.018	1.012	....	....	....	....	....	....	....
0.70	1.022	1.014	....	....	....	....	....	....	....
0.75	1.027	1.016	1.010	....	....	....	....	....	....
0.80	1.037	1.019	1.012	....	....	....	....	....	....
0.90	*1.107*	1.028	1.016	1.010	....	....	....	....	....
1.00	....	1.052	1.022	1.013	....	....	....	....	....
1.10	....	....	1.033	1.016	1.010	....	....	....	....
1.20	....	....	1.062	1.022	1.012	....	....	....	....
1.30	....	....	....	1.032	1.015	....	....	....	....
1.40	....	....	....	1.053	1.019	1.010	....	....	....
1.50	....	....	....	....	1.025	1.012	....	....	....
1.60	....	....	....	....	1.037	1.015	....	....	....
1.70	....	....	....	....	1.067	1.018	....	....	....
1.80	....	....	....	....	....	1.024	1.011	....	....
1.90	....	....	....	....	....	1.033	1.013	....	....
2.00	....	....	....	....	....	1.051	1.015	....	....
2.20	....	....	....	....	....	....	1.022	....	....
2.40	....	....	....	....	....	....	1.043	1.012	....
2.60	....	....	....	....	....	....	....	1.016	....
2.80	....	....	....	....	....	....	....	1.024	....
3.00	....	....	....	....	....	....	....	1.046	1.010

If β<0.5 or ψ> 0.80 use the mean logarithmic temperature difference for parallel flow
values in italics : $2<\gamma<4$

Fig. A.25

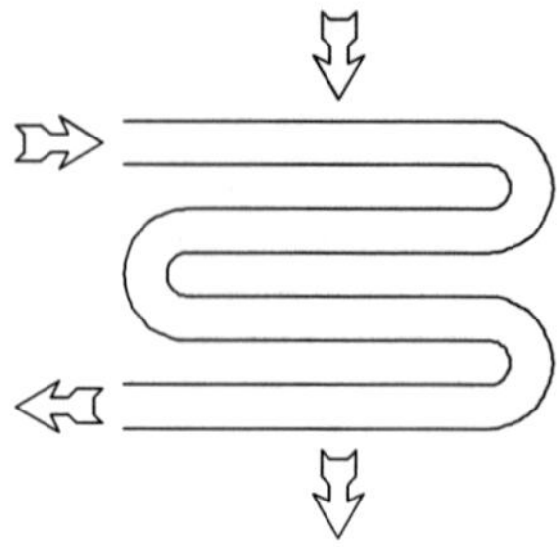

Table A.25 – Coils – parallel flow – 4 sections – Corrective factor χ_p

β \ ψ	0.12	0.16	0.20	0.24	0.28	0.32	0.36	0.40	0.44
0.10	*1.011*	....	....	....	....	....	....	....	....
0.15	....	*1.014*	....	....	....	....	....	....	....
0.20	....	....	*1.015*	....	....	....	....	....	....
0.25	....	....	....	*1.015*	....	....	....	....	....
0.30	....	....	....	*1.036*	*1.014*	....	....	....	....
0.35	....	....	....	....	*1.025*	1.013	....	....	....
0.40	....	....	....	....	....	*1.020*	1.011	....	....
0.45	....	....	....	....	....	*1.039*	1.016	1.010	....
0.50	....	....	....	....	....	....	*1.024*	1.013	....
0.55	....	....	....	....	....	....	*1.050*	1.017	1.010
0.60	....	....	....	....	....	....	....	1.025	1.013
0.65	....	....	....	....	....	....	....	*1.047*	1.016
0.70	....	....	....	....	....	....	....	....	1.023
0.75	....	....	....	....	....	....	....	....	*1.036*

If $\beta > 0.75$ or $\psi < 0.12$ no solution is possible in this field
Values in italics : $2 < \gamma < 4$

Table A.25 – (continued)

β	ψ 0.48	0.52	0.56	0.60	0.64	0.68	0.72	0.76
0.7	1.012	....	....	....	....	....	....	....
0.8	1.019	1.010	....	....	....	....	....	....
0.9	*1.042*	1.015	....	....	....	....	....	....
1.0	....	1.025	1.012	....	....	....	....	....
1.1	....	....	1.017	....	....	....	....	....
1.2	....	....	1.027	1.012	....	....	....	....
1.3	....	....	....	1.016	....	....	....	....
1.4	....	....	....	1.023	1.010	....	....	....
1.5	....	....	....	....	1.013	....	....	....
1.6	....	....	....	....	1.017	....	....	....
1.7	....	....	....	....	1.025	1.010	....	....
1.8	....	....	....	....	....	1.012	....	....
1.9	....	....	....	....	....	1.015	....	....
2.0	....	....	....	....	....	1.020	....	....
2.1	....	....	....	....	....	....	....	....
2.2	....	....	....	....	....	....	1.011	....
2.3	....	....	....	....	....	....	1.014	....
2.4	....	....	....	....	....	....	1.017	....
2.5	....	....	....	....	....	1.015	1.020	....
2.6	....	....	....	....	....	1.036	1.014	....
2.7	....	....	....	....	....	....	1.025	1.010
2.8	....	....	....	....	....	....	....	1.011
2.9	....	....	....	....	....	....	....	1.013
3.0	....	....	....	....	....	....	....	1.016

If β<0.7 or ψ> 0.76 use the mean logarithmic temperature difference for parallel flow
Values in italics : 2<γ <4

A.3.2 Coils with Fluids in Counter Flow (Fig. 2.9)

Fig. A.26

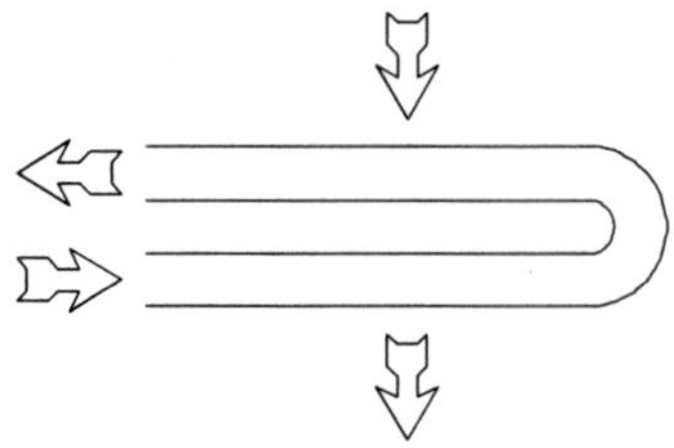

Table A.26 – Coils – counter flow – 2 sections – Corrective factor χ_c

β \ ψ	0.04	0.08	0.12	0.16	0.20	0.24	0.28	0.32	0.36	0.40	0.44
0.1	*0.938*	*0.966*	*0.976*	0.983	0.987	....	....	....	....	....	....
0.2	**0.840**	*0.917*	*0.946*	*0.962*	0.971	0.978	0.983	0.986	0.989	....	....
0.3	....	*0.845*	*0.903*	*0.932*	*0.950*	0.963	0.971	0.977	0.982	0.986	0.989
0.4	....	**0.728**	*0.841*	*0.894*	*0.924*	0.944	0.957	0.968	0.974	0.980	0.984
0.5	....	....	**0.746**	*0.839*	*0.888*	*0.919*	0.939	0.954	0.965	0.972	0.979
0.6	....	....	....	*0.754*	*0.837*	*0.885*	0.917	0.938	0.953	0.964	0.972
0.7	....	....	....	**0.604**	*0.761*	*0.839*	*0.886*	0.917	0.938	0.954	0.965
0.8	....	....	....	....	**0.630**	*0.770*	*0.844*	0.889	0.919	0.941	0.956
0.9	....	....	....	....	....	**0.653**	*0.781*	*0.851*	0.895	0.924	0.945
1.0	....	....	....	....	....	....	*0.679*	*0.796*	*0.862*	0.903	0.931
1.1	....	....	....	....	....	....	....	*0.708*	*0.815*	0.876	0.913
1.2	....	....	....	....	....	....	....	**0.527**	*0.740*	*0.835*	0.891
1.3	....	....	....	....	....	....	....	....	**0.601**	*0.776*	0.859
1.4	....	....	....	....	....	....	....	....	....	*0.672*	*0.813*
1.5	....	....	....	....	....	....	....	....	....	....	*0.738*
1.6	....	....	....	....	....	....	....	....	....	....	**0.586**

If β >1.6 no solution is possible in this field

Values in italics : 2< γ <4 ; **Values in bold type : 4< γ <6**

Table A.26 – (continued)

	ψ								
β	0.48	0.52	0.56	0.60	0.64	0.68	0.72	0.76	0.80
0.4	0.988	....	....	....	....	....	....	....	....
0.5	0.984	0.987	....	....	....	....	....	....	....
0.6	0.979	0.984	0.988	....	....	....	....	....	....
0.7	0.973	0.980	0.985	0.989	....	....	....	....	....
0.8	0.968	0.975	0.981	0.986	....	....	....	....	....
0.9	0.960	0.970	0.978	0.984	0.988	....	....	....	....
1.0	0.950	0.964	0.973	0.981	0.986	....	....	....	....
1.1	0.939	0.956	0.969	0.977	0.984	0.989	....	....	....
1.2	0.925	0.947	0.963	0.974	0.982	0.987	....	....	....
1.3	0.907	0.936	0.956	0.969	0.979	0.986	....	....	....
1.4	0.883	0.923	0.948	0.965	0.976	0.984	0.989	....	....
1.5	0.849	0.905	0.938	0.959	0.972	0.981	0.988	....	....
1.6	*0.798*	0.882	0.926	0.952	0.969	0.979	0.986	....	....
1.7	*0.708*	0.848	0.910	0.944	0.964	0.976	0.985	....	....
1.8	**0.479**	*0.796*	0.889	0.933	0.959	0.973	0.983	....	....
1.9	....	*0.701*	0.859	0.920	0.952	0.970	0.981	0.989	....
2.0	....	....	0.813	0.904	0.945	0.967	0.979	0.988	....
2.2	....	....	**0.477**	0.847	0.923	0.957	0.975	0.985	....
2.4	....	....	....	*0.662*	0.887	0.944	0.969	0.982	....
2.6	....	....	....	....	0.801	0.923	0.961	0.979	0.989
2.8	....	....	....	....	....	0.886	0.950	0.975	0.987
3.0	....	....	....	....	....	0.788	0.934	0.969	0.985

If $\beta < 0.4$ or $\psi > 0.8$ use the mean logarithmic temperature difference for counter flow

Values in italics : $2<\gamma<4$; **Values in bold type : $4<\gamma<6$**

Fig. A.27

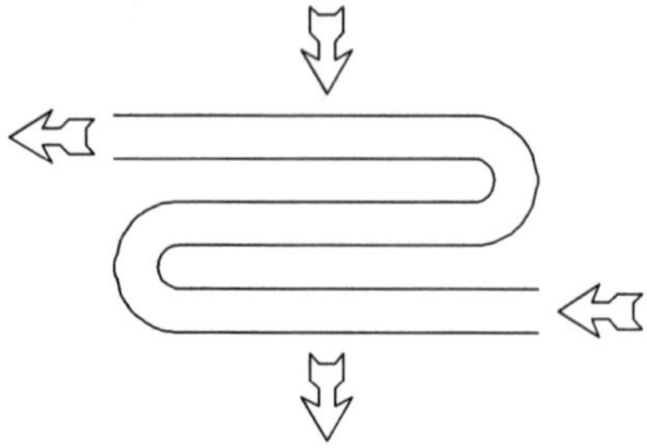

Table A.27 – Coils – counter flow – 3 sections – Corrective factor χ_c

β	ψ 0.04	0.08	0.12	0.16	0.20	0.24	0.28	0.32	0.36	0.40	0.44
0.1	*0.975*	*0.985*	....	....	....	....	....	....	....	....	....
0.2	*0.942*	*0.967*	*0.977*	*0.983*	0.987	....	....	....	....	....	....
0.3	**0.890**	*0.940*	*0.961*	*0.971*	0.979	0.984	0.987	....	....	....	....
0.4	**0.806**	*0.901*	*0.937*	*0.956*	*0.968*	0.975	0.981	0.985	0.989	....	....
0.5	....	**0.842**	*0.904*	*0.935*	*0.953*	0.966	0.974	0.980	0.984	0.988	....
0.6	....	**0.739**	**0.853**	*0.904*	*0.933*	*0.952*	0.965	0.973	0.979	0.984	0.988
0.7	....	....	**0.767**	*0.857*	*0.905*	*0.933*	*0.951*	0.964	0.973	0.979	0.984
0.8	....	....	....	**0.779**	*0.861*	*0.906*	*0.934*	0.952	0.965	0.974	0.980
0.9	....	....	....	....	**0.787**	*0.865*	*0.909*	*0.936*	0.954	0.967	0.975
1.0	....	....	....	....	....	*0.795*	*0.871*	*0.914*	0.941	0.958	0.969
1.1	....	....	....	....	....	**0.645**	*0.806*	*0.879*	*0.920*	0.946	0.962
1.2	....	....	....	....	....	....	**0.664**	*0.820*	*0.891*	0.929	0.952
1.3	....	....	....	....	....	....	....	**0.692**	*0.839*	*0.903*	0.938
1.4	....	....	....	....	....	....	....	....	**0.731**	*0.862*	0.918
1.5	....	....	....	....	....	....	....	....	....	*0.778*	*0.886*
1.6	....	....	....	....	....	....	....	....	....	....	*0.827*
1.7	....	....	....	....	....	....	....	....	....	....	**0.668**

If β> 1.7 no solution is possible in this field

Values in italics : 2<γ <4 ; **Values in bold type : 4<γ<6**

Table A.27 – (continued)

β	ψ 0.48	0.52	0.56	0.60	0.64	0.68	0.72	0.76
0.7	0.988	....	....	....	....	....	....	....
0.8	0.985	0.989	....	....	....	....	....	....
0.9	0.982	0.987	....	....	....	....	....	....
1.0	0.978	0.984	0.988	....	....	....	....	....
1.1	0.973	0.980	0.986	....	....	....	....	....
1.2	0.967	0.976	0.983	0.988	....	....	....	....
1.3	0.959	0.971	0.980	0.986	....	....	....	....
1.4	0.947	0.966	0.976	0.984	0.989	....	....	....
1.5	0.932	0.958	0.972	0.981	0.987	....	....	....
1.6	0.910	0.947	0.967	0.978	0.986	....	....	....
1.7	*0.871*	0.931	0.960	0.974	0.984	0.989	....	....
1.8	*0.789*	0.908	0.949	0.969	0.981	0.988	....	....
1.9	....	*0.866*	0.936	0.964	0.978	0.986	....	....
2.0	....	*0.761*	0.915	0.956	0.975	0.985	....	....
2.2	....	....	*0.778*	0.930	0.965	0.980	0.988	....
2.4	....	....	....	*0.844*	0.947	0.974	0.986	....
2.6	....	....	....	....	0.906	0.965	0.982	....
2.8	....	....	....	....	....	0.947	0.977	0.988
3.0	....	....	....	....	....	0.898	0.969	0.986

If $\beta < 0.7$ or $\psi > 0.76$ use the mean logarithmic temperature difference for counter flow

Values in italics : $2 < \gamma < 4$

Fig. A.28

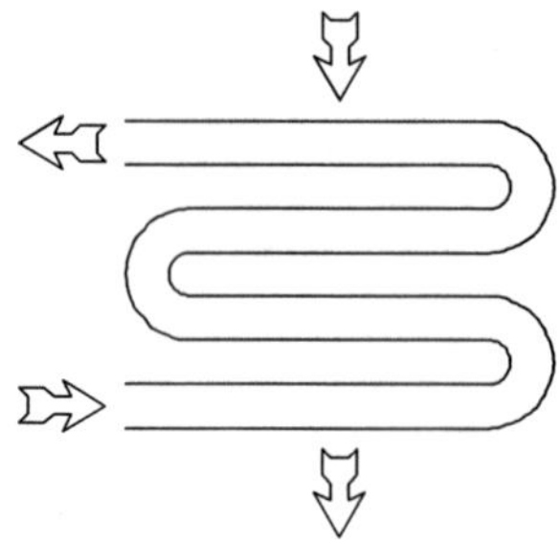

Table A.28 – Coils – counter flow – 4 sections – Corrective factor χ_c

β	ψ 0.04	0.08	0.12	0.16	0.20	0.24	0.28	0.32	0.36	0.40	0.44
0.1	*0.986*	....	....	....	....	....	....	....	....	....	....
0.2	*0.969*	*0.981*	*0.987*	....	....	....	....	....	....	....	....
0.3	**0.942**	*0.967*	*0.978*	*0.984*	0.988	....	....	....	....	....	....
0.4	**0.902**	*0.947*	*0.966*	*0.975*	*0.982*	0.986	0.989	....	....	....	....
0.5	....	**0.917**	*0.947*	*0.964*	*0.973*	0.980	0.985	0.988	....	....	....
0.6	....	**0.869**	*0.921*	*0.947*	*0.963*	*0.972*	0.979	0.984	0.988	....	....
0.7	....	....	**0.879**	*0.923*	*0.947*	*0.963*	0.972	0.979	0.984	0.988	....
0.8	....	....	**0.804**	**0.883**	*0.924*	*0.947*	*0.963*	0.973	0.980	0.985	0.989
0.9	....	....	....	**0.812**	*0.886*	*0.926*	*0.949*	0.965	0.974	0.981	0.986
1.0	....	....	....	....	**0.816**	*0.890*	*0.929*	*0.951*	0.967	0.976	0.982
1.1	....	....	....	....	....	**0.819**	*0.894*	*0.932*	*0.955*	0.969	0.978
1.2	....	....	....	....	....	....	**0.825**	*0.901*	*0.938*	0.960	0.972
1.3	....	....	....	....	....	....	....	**0.837**	*0.910*	*0.946*	0.965
1.4	....	....	....	....	....	....	....	....	*0.854*	*0.922*	0.953
1.5	....	....	....	....	....	....	....	....	....	*0.877*	*0.935*
1.6	....	....	....	....	....	....	....	....	....	**0.740**	*0.903*
1.7	....	....	....	....	....	....	....	....	....	....	*0.820*

If β> 1.7 no solution is possible in this field
Values in italics : 2<γ <4 ; **Values in bold type : 4< γ <6**

Table A.28 – (continued)

β	ψ 0.48	0.52	0.56	0.60	0.64	0.68	0.72
1.0	0.987	…….	…….	…….	…….	…….	…….
1.1	0.984	0.989	…….	…….	…….	…….	…….
1.2	0.981	0.986	…….	…….	…….	…….	…….
1.3	0.976	0.984	0.989	…….	…….	…….	…….
1.4	0.970	0.980	0.986	…….	…….	…….	…….
1.5	0.962	0.975	0.984	0.989	…….	…….	…….
1.6	0.948	0.969	0.981	0.987	…….	…….	…….
1.7	*0.927*	0.961	0.976	0.985	…….	…….	…….
1.8	*0.882*	0.947	0.971	0.982	0.989	…….	…….
1.9	…….	*0.923*	0.963	0.979	0.987	…….	…….
2.0	…….	*0.865*	0.951	0.974	0.985	…….	…….
2.2	…….	…….	*0.873*	0.959	0.979	0.989	…….
2.4	…….	…….	…….	*0.909*	0.969	0.985	…….
2.6	…….	…….	…….	…….	0.946	0.979	…….
2.8	…….	…….	…….	…….	…….	0.969	0.987
3.0	…….	…….	…….	…….	…….	0.940	0.982

If $\beta < 1$ or $\psi > 0.72$ use the mean logarithmic temperature difference for counter flow

Values in italics : $2 < \gamma < 4$

Fig. A.29

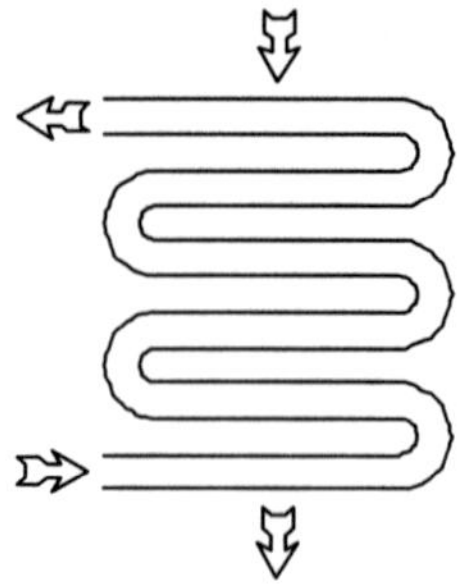

Table A.29 – Coils – counter flow – 6 sections – Corrective factor χ_c

	ψ										
β	0.04	0.08	0.12	0.16	0.20	0.24	0.28	0.32	0.36	0.40	0.44
0.2	*0.986*	....	....	....	....	....	....	....	....	....	....
0.3	**0.975**	*0.985*	....	....	....	....	....	....	....	....	....
0.4	**0.959**	*0.977*	*0.985*	*0.989*	....	....	....	....	....	....	....
0.5	**0.934**	*0.965*	*0.977*	*0.984*	*0.988*	....	....	....	....	....	....
0.6	....	**0.946**	*0.966*	*0.977*	*0.983*	*0.988*	....	....	....	....	....
0.7	....	**0.914**	**0.949**	*0.967*	*0.977*	*0.983*	0.988	....	....	....	....
0.8	....	....	**0.919**	*0.950*	*0.967*	*0.977*	*0.983*	0.988	....	....	....
0.9	....	....	....	**0.922**	*0.951*	*0.968*	*0.977*	0.984	0.988	....	....
1.0	....	....	....	....	**0.922**	*0.952*	*0.969*	*0.978*	0.985	0.989	....
1.1	....	....	....	....	**0.856**	**0.923**	*0.954*	*0.970*	0.980	0.986	....
1.2	....	....	....	....	....	**0.851**	*0.925*	*0.956*	*0.972*	0.982	0.987
1.3	....	....	....	....	....	....	**0.845**	*0.930*	*0.961*	*0.975*	0.984
1.4	....	....	....	....	....	....	....	**0.846**	*0.936*	*0.966*	0.979
1.5	....	....	....	....	....	....	....	....	**0.861**	*0.946*	*0.971*
1.6	....	....	....	....	....	....	....	....	....	**0.889**	*0.957*
1.7	....	....	....	....	....	....	....	....	....	....	*0.920*

If $\beta > 1.7$ no solution is possible in this field

Values in italics : $2<\gamma<4$; **Values in bold type : $4<\gamma<6$**

Table A.29 – (continued)

β	ψ 0.48	0.52	0.56	0.60	0.64	0.68
1.3	0.989	....	....	....	....	....
1.4	0.986	....	....	....	....	....
1.5	0.982	0.989	....	....	....	....
1.6	0.977	0.986	....	....	....	....
1.7	*0.968*	0.982	0.989	....	....	....
1.8	*0.947*	0.976	0.987	....	....	....
1.9	**0.851**	*0.966*	0.983	....	....	....
2.0	....	*0.939*	0.977	0.988	....	....
2.2	....	....	*0.943*	0.981	....	....
2.4	....	....	....	*0.959*	0.986	....
2.6	....	....	....	....	0.975	....
2.8	....	....	....	....	....	0.985
3.0	....	....	....	....	....	0.972

If $\beta<1.3$ or $\psi>0.68$ use the mean logarithmic temperature difference for counter flow

Values in italics : $2<\gamma<4$; **Value in bold type : $4<\gamma<6$**

Table A.30 – Coils – counter flow – 8 sections – Corrective factor χ_c

β \ ψ	0.04	0.08	0.12	0.16	0.20	0.24	0.28	0.32	0.36	0.40	0.44
0.3	**0.986**	…….	…….	…….	…….	…….	…….	…….	…….	…….	…….
0.4	**0.977**	*0.987*	…….	…….	…….	…….	…….	…….	…….	…….	…….
0.5	**0.964**	*0.980*	*0.987*	…….	…….	…….	…….	…….	…….	…….	…….
0.6	…….	**0.969**	*0.981*	*0.987*	…….	…….	…….	…….	…….	…….	…….
0.7	…….	**0.953**	*0.971*	*0.981*	*0.987*	…….	…….	…….	…….	…….	…….
0.8	…….	…….	**0.956**	*0.972*	*0.981*	*0.987*	…….	…….	…….	…….	…….
0.9	…….	…….	**0.925**	**0.957**	*0.972*	*0.981*	*0.987*	…….	…….	…….	…….
1.0	…….	…….	…….	**0.925**	**0.957**	*0.973*	*0.982*	*0.988*	…….	…….	…….
1.1	…….	…….	…….	…….	**0.922**	*0.957*	*0.974*	*0.983*	0.988	…….	…….
1.2	…….	…….	…….	…….	…….	**0.919**	*0.958*	*0.975*	*0.984*	0.989	…….
1.3	…….	…….	…….	…….	…….	…….	**0.915**	*0.961*	*0.977*	*0.986*	…….
1.4	…….	…….	…….	…….	…….	…….	…….	**0.916**	*0.964*	*0.980*	0.988
1.5	…….	…….	…….	…….	…….	…….	…….	…….	**0.923**	*0.969*	*0.983*
1.6	…….	…….	…….	…….	…….	…….	…….	…….	…….	**0.938**	*0.975*
1.7	…….	…….	…….	…….	…….	…….	…….	…….	…….	…….	*0.955*

β \ ψ	0.48	0.52	0.56	0.60	0.64	0.68
1.6	0.987	…….	…….	…….	…….	…….
1.7	*0.981*	…….	…….	…….	…….	…….
1.8	*0.969*	0.986	…….	…….	…….	…….
1.9	**0.917**	*0.980*	…….	…….	…….	…….
2.0	…….	*0.966*	0.987	…….	…….	…….
2.2	…….	…….	*0.968*	0.989	…….	…….
2.4	…….	…….	…….	0.976	…….	…….
2.6	…….	…….	…….	…….	0.986	…….
2.8	…….	…….	…….	…….	…….	…….
3.0	…….	…….	…….	…….	…….	0.984

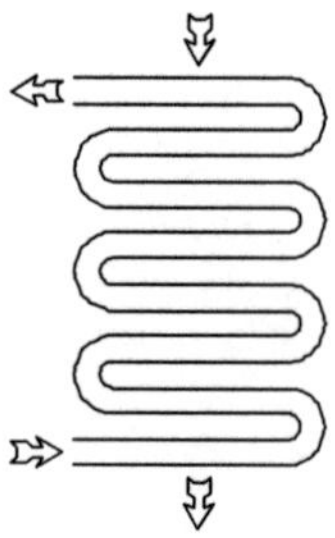

Fig. A.30

If $\psi > 0.68$ use the mean logarithmic temperature difference for counter flow *Values in italics : 2< γ <4 ;* **Values in bold type : 4< γ <6**

Table A.31 – Coils – counter flow – 10 sections – Corrective factor χ_c

β	ψ 0.04	0.08	0.12	0.16	0.20	0.24	0.28	0.32	0.36	0.40	0.44
0.4	**0.985**	....	....	....	....	....	....	....	....	....	....
0.5	**0.977**	*0.987*	....	....	....	....	....	....	....	....	....
0.6	....	**0.980**	*0.988*	....	....	....	....	....	....	....	....
0.7	....	**0.969**	*0.982*	*0.988*	....	....	....	....	....	....	....
0.8	....	....	**0.971**	*0.982*	*0.988*	....	....	....	....	....	....
0.9	....	....	**0.953**	**0.972**	*0.982*	*0.988*	....	....	....	....	....
1.0	....	....	....	**0.952**	**0.972**	*0.983*	*0.988*	....	....	....	....
1.1	....	....	....	....	**0.950**	*0.972*	*0.983*	*0.989*	....	....	....
1.2	....	....	....	....	....	**0.948**	*0.973*	*0.984*	....	....	....
1.3	....	....	....	....	....	....	**0.947**	*0.974*	*0.985*	....	....
1.4	....	....	....	....	....	....	....	**0.947**	*0.977*	*0.987*	....
1.5	....	....	....	....	....	....	....	....	**0.950**	*0.980*	*0.989*
1.6	....	....	....	....	....	....	....	....	....	*0.960*	*0.984*
1.7	....	....	....	....	....	....	......				*0.971*

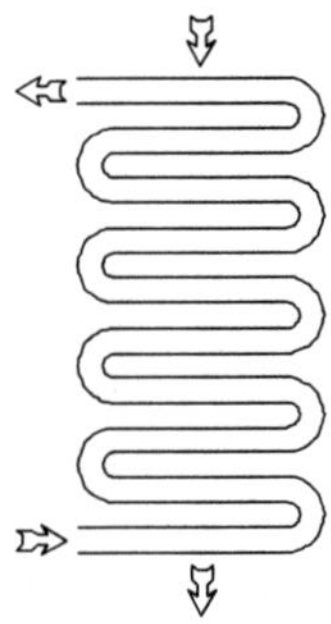

Fig. A.31

β	ψ 0.48	0.52	0.56	0.60
1.7	*0.988*	....	....	....
1.8	*0.980*	....	....	....
1.9	**0.947**	0.987	....	....
2.0	....	*0.977*	....	....
2.2	....	....	*0.979*	....
2.4	....	....	....	0.985

If $\psi > 0.6$ use the mean logarithmic temper ature difference for counter flow *Values in ital ics : $2<\gamma<4$:* **Values in bold type : $4<\gamma<6$**

A.4 Tube Banks with Several Passages of External Fluid

A.4.1 Tube Banks with Fluids in Parallel Flow (Fig. 2.10)

Fig. A.32

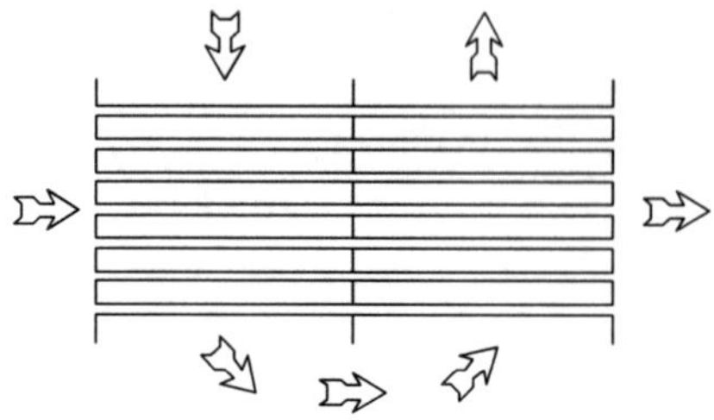

Table A.32 – Tube bank – parallel flow – 2 passages of external fluid Corrective factor χ_p

β \ ψ	0.12	0.16	0.20	0.24	0.28	0.32	0.36	0.40	0.44
0.10	*1.060*	*1.031*	1.020	1.014	1.010	....	....	....	....
0.15	....	*1.075*	*1.039*	1.025	1.018	1.013	....	....	....
0.20	....	....	*1.080*	1.043	1.028	1.020	1.014	1.011	....
0.25	....	....	....	*1.076*	1.043	1.029	1.020	1-015	1.011
0.30	....	....	....	*1.181*	1.068	1.041	1.027	1.019	1.014
0.35	....	....	....	....	*1.124*	1.060	1.037	1.025	1.018
0.40	....	....	....	....	....	1.092	1.051	1.033	1.022
0.45	....	....	....	....	....	*1.174*	1.071	1.042	1.028
0.50	....	....	....	....	....	....	1.107	1.055	1.035
0.55	....	....	....	....	....	....	....	1.075	1.044
0.60	....	....	....	....	....	....	....	1.108	1.055
0.65	....	....	....	....	....	....	....	....	1.071
0.70	....	....	....	....	....	....	....	....	1.096
0.75	....	....	....	....	....	....	....	....	1.131

If $\beta > 0.75$ or $\psi < 0.12$ no solution is possible in this field
Values in italics : $2<\gamma<4$

Table A.32 – (continued)

β	ψ 0.48	0.52	0.56	0.60	0.64	0.68	0.72	0.76	0.80
0.3	1.010	....	....	....	....	....	....	....	....
0.4	1.016	1.012	....	....	....	....	....	....	....
0.5	1.023	1.017	1.012	....	....	....	....	....	....
0.6	1.034	1.023	1.016	1.011	....	....	....	....	....
0.7	1.050	1.031	1.021	1.014	....	....	....	....	....
0.8	1.079	1.044	1.027	1.018	1.012	....	....	....	....
0.9	....	1.062	1.036	1.022	1.015	....	....	....	....
1.0	....	1.094	1.048	1.028	1.018	1.011	....	....	....
1.1	....	....	1.066	1.036	1.021	1.013	....	....	....
1.2	....	....	1.085	1.046	1.026	1.016	....	....	....
1.3	....	....	....	1.060	1.032	1.019	1.011	....	....
1.4	....	....	....	1.073	1.039	1.022	1.013	....	....
1.5	....	....	....	....	1.049	1.026	1.015	....	....
1.6	....	....	....	....	1.060	1.031	1.017	....	....
1.7	....	....	....	....	1.030	1.036	1.019	1.011	....
1.8	....	....	....	....	....	1.044	1.022	1.012	....
1.9	....	....	....	....	....	1.051	1.026	1.013	....
2.0	....	....	....	....	....	1.045	1.029	1.015	....
2.1	....	....	....	....	....	....	1.034	1.017	....
2.2	....	....	....	....	....	....	1.039	1.019	....
2.3	....	....	....	....	....	....	1.043	1.021	1.010
2.4	....	....	....	....	....	....	1.039	1.023	1.011
2.5	....	....	....	....	....	....	....	1.026	1.012
2.6	....	....	....	....	....	....	....	1.029	1.013
2.7	....	....	....	....	....	....	....	1.032	1.015
2.8	....	....	....	....	....	....	....	1.035	1.016
2.9	....	....	....	....	....	....	....	1.035	1.017
3.0	....	....	....	....	....	....	....	1.012	1.019

If β< 0.3 or ψ>0.8 use the mean logarithmic temperature difference for parallel flow

Fig. A.33

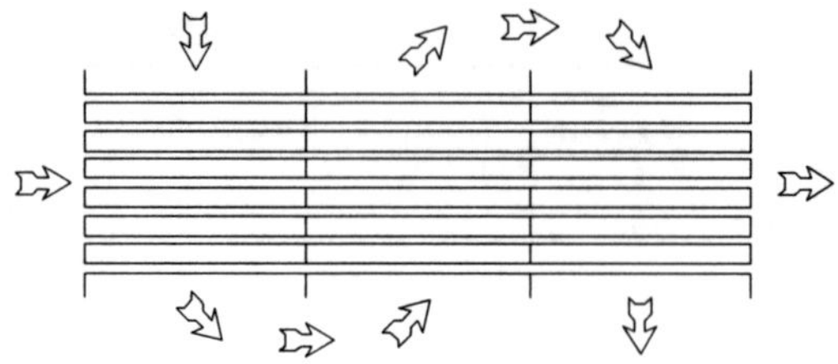

Table A.33 – Tube banks – parallel flow – 3 passages of external fluid Corrective factor χ_p

β \ ψ	0.12	0.16	0.20	0.24	0.28	0.32	0.36	0.40	0.44
0.10	*1.024*	*1.012*	....	....	....	....	....	....	....
0.15	....	*1.031*	*1.016*	1.011	....	....	....	....	....
0.20	....	....	*1.034*	1.018	1.012	....	....	....	....
0.25	....	....	....	*1.033*	1.019	1.012	....	....	....
0.30	....	....	....	*1.085*	*1.030*	1.018	1.012	....	....
0.35	....	....	....	....	*1.057*	1.026	1.016	1.011	....
0.40	....	....	....	....	....	*1.042*	1.022	1.014	....
0.45	....	....	....	....	....	*1.093*	1.032	1.019	1.012
0.50	....	....	....	....	....	....	*1.051*	1.025	1.016
0.55	....	....	....	....	....	....	*1.131*	1.035	1.020
0.60	....	....	....	....	....	....	....	1.054	1.026
0.65	....	....	....	....	....	....	....	*1.123*	1.034
0.70	....	....	....	....	....	....	....	....	1.049
0.75	....	....	....	....	....	....	....	....	*1.087*

If $\beta > 0.75$ or $\psi < 0.12$ no solution is possible in this field
Values in italics : $2 < \gamma < 4$

Table A.33 – (continued)

β	ψ 0.48	0.52	0.56	0.60	0.64	0.68	0.72	0.76
0.5	1.010	....	....	....	....	....	....	....
0.6	1.016	1.010	....	....	....	....	....	....
0.7	1.024	1.014	....	....	....	....	....	....
0.8	1.040	1.020	1.012	....	....	....	....	....
0.9	*1.120*	1.031	1.017	1.010	....	....	....	....
1.0	....	1.058	1.023	1.013	....	....	....	....
1.1	....	....	1.035	1.017	1.010	....	....	....
1.2	....	....	1.067	1.023	1.012	....	....	....
1.3	....	....	....	1.034	1.016	....	....	....
1.4	....	....	....	1.057	1.020	1.010	....	....
1.5	....	....	....	....	1.027	1.013	....	....
1.6	....	....	....	....	1.039	1.015	....	....
1.7	....	....	....	....	1.072	1.019	....	....
1.8	....	....	....	....	....	1.025	1.011	....
1.9	....	....	....	....	....	1.034	1.013	....
2.0	....	....	....	....	....	1.054	1.015	....
2.1	....	....	....	....	....	1.166	1.019	....
2.2	....	....	....	....	....	....	1.023	....
2.3	....	....	....	....	....	....	1.031	1.010
2.4	....	....	....	....	....	....	1.045	1.012
2.5	....	....	....	....	....	....	1.087	1.014
2.6	....	....	....	....	....	....	....	1.017
2.7	....	....	....	....	....	....	....	1.020
2.8	....	....	....	....	....	....	....	1.025
2.9	....	....	....	....	....	....	....	1.033
3.0	....	....	....	....	....	....	....	1.048

If $\beta < 0.5$ or $\psi > 0.76$ use the mean logarithmic temperature difference for parallel flow

Value in italics : $2 < \gamma < 4$

A.4.2 Tube Banks with Fluids in Counter Flow (Fig. 2.11)

Fig. A.34

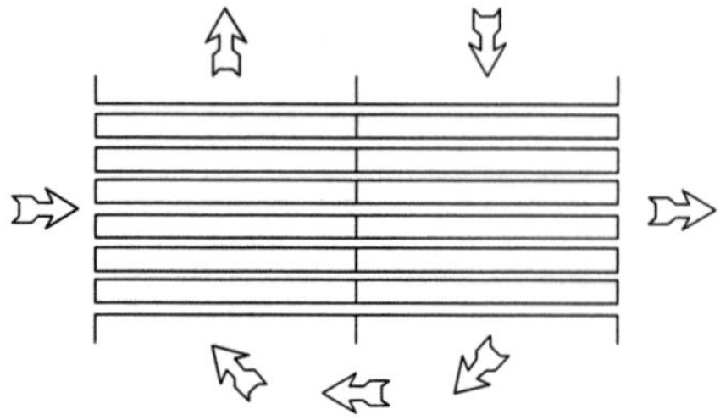

Table A.34 – Tube banks – counter flow – 2 passages of external fluid – Corrective factor χ_c

β \ ψ	0.04	0.08	0.12	0.16	0.20	0.24	0.28	0.32	0.36	0.40	0.44
0.1	*0.965*	*0.977*	*0.983*	0.987	....	....	....	....	....	....	....
0.2	**0.920**	*0.948*	*0.963*	*0.971*	0.977	0.982	0.986	0.988	....	....	....
0.3	**0.865**	*0.912*	*0.936*	*0.952*	0.963	0.970	0.976	0.981	0.985	0.988	....
0.4	**0.799**	*0.867*	*0.904*	*0.928*	*0.945*	0.957	0.966	0.972	0.978	0.983	0.986
0.5	....	**0.811**	*0.863*	*0.898*	*0.922*	*0.940*	0.952	0.963	0.970	0.976	0.981
0.6	....	**0.743**	**0.812**	*0.860*	*0.893*	*0.919*	0.936	0.950	0.961	0.969	0.976
0.7	....	....	**0.749**	*0.811*	*0.857*	*0.891*	*0.917*	0.935	0.950	0.961	0.969
0.8	....	....	....	**0.751**	*0.811*	*0.856*	*0.891*	0.917	0.936	0.951	0.963
0.9	....	....	....	....	**0.751**	*0.812*	*0.858*	*0.893*	0.919	0.939	0.954
1.0	....	....	....	....	**0.673**	**0.754**	*0.816*	*0.863*	0.898	0.924	0.943
1.1	....	....	....	....	....	**0.674**	*0.759*	*0.823*	*0.870*	0.905	0.930
1.2	....	....	....	....	....	....	**0.679**	*0.768*	*0.834*	*0.881*	0.914
1.3	....	....	....	....	....	....	....	**0.689**	*0.783*	*0.848*	0.893
1.4	....	....	....	....	....	....	....	....	**0.707**	*0.803*	*0.865*
1.5	....	....	....	....	....	....	....	....	....	*0.735*	*0.827*
1.6	....	....	....	....	....	....	....	....	....	....	*0.771*
1.7	....	....	....	....	....	....	....	....	....	....	**0.674**

If $\beta > 1.7$ no solution is possible in this field

Values in italics : $2<\gamma<4$; **Values in bold type : $4<\gamma<6$**

Table A.34 – (continued)

β \ ψ	0.48	0.52	0.56	0.60	0.64	0.68	0.72	0.76	0.80
0.4	0.989	....	....	....	....	....	....	....	....
0.5	0.985	0.989	....	....	....	....	....	....	....
0.6	0.981	0.985	0.989	....	....	....	....	....	....
0.7	0.976	0.982	0.986	0.989	....	....	....	....	....
0.8	0.971	0.978	0.983	0.987	....	....	....	....	....
0.9	0.965	0.973	0.980	0.985	0.989	....	....	....	....
1.0	0.957	0.969	0.976	0.982	0.987	....	....	....	....
1.1	0.948	0.962	0.972	0.979	0.985	....	....	....	....
1.2	0.937	0.955	0.968	0.976	0.983	0.988	....	....	....
1.3	0.924	0.946	0.962	0.972	0.981	0.986	....	....	....
1.4	0.908	0.935	0.955	0.969	0.978	0.985	....	....	....
1.5	0.885	0.922	0.947	0.964	0.975	0.983	0.988	....	....
1.6	*0.855*	0.906	0.937	0.958	0.971	0.981	0.987	....	....
1.7	*0.811*	0.883	0.925	0.951	0.968	0.978	0.986	....	....
1.8	*0.741*	0.852	0.909	0.943	0.963	0.976	0.984	....	....
1.9	....	*0.806*	0.890	0.932	0.958	0.973	0.983	0.989	....
2.0	....	*0.723*	0.861	0.919	0.951	0.969	0.981	0.988	....
2.1	....	....	*0.817*	0.903	0.944	0.966	0.979	0.987	....
2.2	....	....	*0.734*	0.880	0.934	0.961	0.976	0.986	....
2.3	....	....	....	0.846	0.922	0.956	0.974	0.985	....
2.4	....	....	....	*0.786*	0.907	0.949	0.971	0.983	....
2.5	....	....	....	....	0.884	0.942	0.969	0.982	....
2.6	....	....	....	....	0.851	0.933	0.965	0.980	0.989
2.7	....	....	....	....	*0.787*	0.921	0.960	0.978	0.988
2.8	....	....	....	....	....	0.905	0.955	0.976	0.987
2.9	....	....	....	....	....	0.882	0.949	0.974	0.986
3.0	....	....	....	....	....	0.842	0.942	0.971	0.986

If $\beta < 0.4$ or $\psi > 0.8$ use the mean logarithmic temperature difference for counter flow
Values in italics : $2 < \gamma < 4$

Fig. A.35

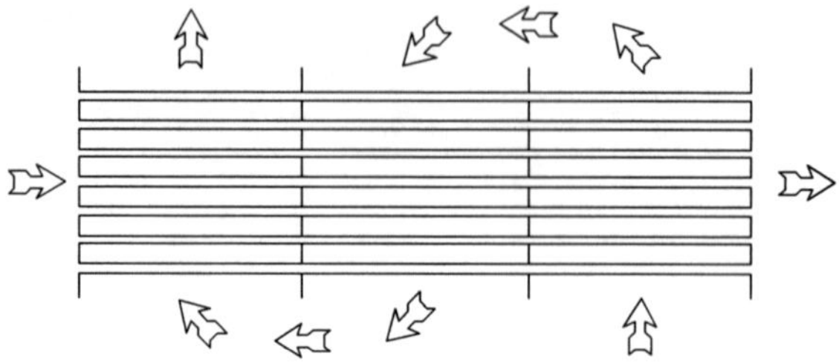

Table A.35 – Tube banks – counter flow – 3 passages of external fluid – Corrective factor χ_c

β \ ψ	0.04	0.08	0.12	0.16	0.20	0.24	0.28	0.32	0.36	0.40	0.44
0.1	*0.978*	*0.986*	....	....	....	....	....	....	....	....	....
0.2	*0.948*	*0.969*	*0.978*	*0.984*	0.988	....	....	....	....	....	....
0.3	**0.907**	*0.946*	*0.964*	*0.973*	0.980	0.984	0.988	....	....	....	....
0.4	**0.847**	*0.914*	*0.943*	*0.959*	*0.969*	0.977	0.982	0.986	0.989	....	....
0.5	....	**0.867**	*0.914*	*0.940*	*0.956*	0.968	0.975	0.981	0.985	0.988	....
0.6	....	**0.799**	*0.873*	*0.914*	*0.939*	*0.955*	0.967	0.974	0.980	0.985	0.988
0.7	....	....	**0.811**	*0.876*	*0.913*	*0.938*	0.954	0.966	0.974	0.980	0.985
0.8	....	....	....	**0.816**	*0.877*	*0.914*	*0.939*	0.955	0.967	0.975	0.981
0.9	....	....	....	**0.724**	**0.820**	*0.879*	*0.917*	*0.941*	0.957	0.969	0.976
1.0	....	....	....	....	**0.727**	*0.825*	*0.883*	*0.920*	0.944	0.960	0.970
1.1	....	....	....	....	....	**0.729**	*0.830*	*0.890*	*0.926*	0.948	0.964
1.2	....	....	....	....	....	....	**0.736**	*0.840*	*0.899*	0.932	0.954
1.3	....	....	....	....	....	....	....	**0.750**	*0.855*	*0.909*	0.941
1.4	....	....	....	....	....	....	....	....	**0.773**	*0.874*	0.922
1.5	....	....	....	....	....	....	....	....	....	*0.806*	*0.894*
1.6	....	....	....	....	....	....	....	....	....	....	*0.844*
1.7	....	....	....	....	....	....	....	....	....	....	**0.725**

If β> 1.7 no solution is possible in this field

Values in italics : 2<γ<4 ; **Values in bold type : 4<γ<6**

Table A.35 – (continued)

	ψ							
β	0.48	0.52	0.56	0.60	0.64	0.68	0.72	0.76
0.7	0.988	....	....	....	....	....	....	....
0.8	0.986	0.989	....	....	....	....	....	....
0.9	0.982	0.987	....	....	....	....	....	....
1.0	0.978	0.984	0.988	....	....	....	....	....
1.1	0.974	0.981	0.986	....	....	....	....	....
1.2	0.968	0.977	0.984	0.988	....	....	....	....
1.3	0.960	0.972	0.981	0.987	....	....	....	....
1.4	0.950	0.967	0.977	0.984	0.989	....	....	....
1.5	0.936	0.959	0.973	0.982	0.988	....	....	....
1.6	0.916	0.949	0.968	0.979	0.986	....	....	....
1.7	*0.881*	0.935	0.961	0.975	0.984	....	....	....
1.8	*0.811*	0.913	0.951	0.970	0.982	0.988	....	....
1.9	....	*0.876*	0.939	0.966	0.979	0.987	....	....
2.0	....	*0.788*	0.919	0.958	0.975	0.985	....	....
2.1	....	....	*0.884*	0.947	0.971	0.983	....	....
2.2	....	....	*0.800*	0.932	0.966	0.981	0.989	....
2.3	....	....	....	0.908	0.958	0.978	0.987	....
2.4	....	....	....	*0.856*	0.949	0.974	0.986	....
2.5	....	....	....	....	0.935	0.970	0.984	....
2.6	....	....	....	....	0.911	0.966	0.982	....
2.7	....	....	....	....	*0.856*	0.958	0.980	....
2.8	....	....	....	....	....	0.948	0.977	0.989
2.9	....	....	....	....	....	0.932	0.974	0.988
3.0	....	....	....	....	....	0.903	0.970	0.986

If $\beta < 0.7$ or $\psi > 0.76$ use the mean logarithmic temperature difference for counter flow
Values in italics : $2 < \gamma < 4$

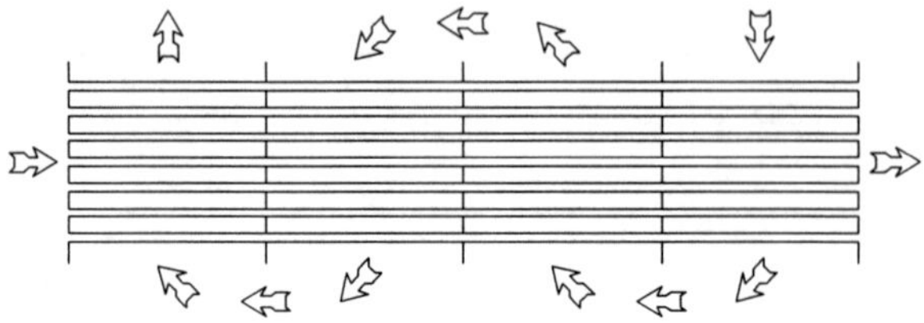

Fig. A.36

Table A.36 – Tube banks – counter flow – 4 passages of external fluid – Corrective factor χ_c

	ψ										
β	0.04	0.08	0.12	0.16	0.20	0.24	0.28	0.32	0.36	0.40	0.44
0.1	*0.986*	....	....	....	....	....	....	....	....	....	....
0.2	*0.969*	*0.981*	*0.987*	....	....	....	....	....	....	....	....
0.3	**0.942**	*0.967*	*0.978*	*0.984*	0.988	....	....	....	....	....	....
0.4	**0.903**	*0.947*	*0.966*	*0.975*	*0.982*	0.986	0.989	....	....	....	....
0.5	....	**0.918**	*0.947*	*0.964*	*0.973*	0.980	0.985	0.988	....	....	....
0.6	....	**0.871**	*0.921*	*0.947*	*0.963*	*0.972*	0.980	0.984	0.988	....	....
0.7	....	....	**0.881**	*0.923*	*0.947*	*0.963*	0.972	0.979	0.985	0.988	....
0.8	....	....	**0.809**	**0.884**	*0.924*	*0.948*	*0.963*	0.973	0.980	0.985	0.989
0.9	....	....	....	**0.816**	*0.887*	*0.926*	*0.949*	0.965	0.974	0.981	0.986
1.0	....	....	....	....	**0.818**	*0.891*	*0.929*	*0.952*	0.967	0.976	0.982
1.1	....	....	....	....	....	**0.821**	*0.894*	*0.932*	*0.955*	0.969	0.978
1.2	....	....	....	....	....	....	**0.827**	*0.901*	*0.938*	0.960	0.972
1.3	....	....	....	....	....	....	....	**0.838**	*0.910*	*0.946*	0.965
1.4	....	....	....	....	....	....	....	....	*0.855*	*0.922*	0.953
1.5	....	....	....	....	....	....	....	....	....	*0.877*	*0.935*
1.6	....	....	....	....	....	....	....	....	....	**0.740**	*0.903*
1.7	....	....	....	....	....	....	....	....	....	....	*0.819*

If $\beta > 1.7$ no solution is possible in this field
Values in italics : $2<\gamma<4$; **Values in bold type : $4<\gamma<6$**

Table A.36 – (continued)

	ψ						
β	0.48	0.52	0.56	0.60	0.64	0.68	0.72
1.0	0.987	....	....	....	....	....	....
1.1	0.984	0.989	....	....	....	....	....
1.2	0.981	0.986	....	....	....	....	....
1.3	0.976	0.984	0.989	....	....	....	....
1.4	0.970	0.980	0.987	....	....	....	....
1.5	0.962	0.975	0.984	0.989	....	....	....
1.6	0.948	0.969	0.981	0.987	....	....	....
1.7	*0.927*	0.961	0.976	0.985	....	....	....
1.8	*0.882*	0.947	0.971	0.983	0.989	....	....
1.9	....	*0.924*	0.964	0.979	0.987	....	....
2.0	....	*0.865*	0.951	0.974	0.985	....	....
2.1	....	....	0.929	0.969	0.983	....	....
2.2	....	....	*0.873*	0.959	0.980	0.989	....
2.3	....	....	....	0.944	0.975	0.987	....
2.4	....	....	....	*0.909*	0.969	0.985	....
2.5	....	....	....	....	0.961	0.982	....
2.6	....	....	....	....	0.946	0.979	....
2.7	....	....	....	....	*0.909*	0.975	0.988
2.8	....	....	....	....	....	0.969	0.987
2.9	....	....	....	....	....	0.959	0.985
3.0	....	....	....	....	....	0.940	0.982

If $\beta < 1.0$ or $\psi > 0.72$ use the mean logarithmic temperature difference for counter flow

Values in italics : $2 < \gamma < 4$

Appendix B
Factor ψ or Corrective Factors for Verification Computation

B.1 Fluids in Parallel Flow or Counter Flow

Table B.1 – Factor ψ for fluids in parallel flow

β / γ	0.05	0.10	0.15	0.20	0.25	0.30	0.35	0.40	0.45	0.50	0.55
0.1	0.951	0.905	0.862	0.820	0.781	0.744	0.710	0.676	0.645	0.615	0.587
0.2	0.951	0.906	0.863	0.822	0.784	0.748	0.714	0.682	0.652	0.624	0.597
0.3	0.952	0.906	0.864	0.824	0.787	0.752	0.719	0.688	0.659	0.632	0.607
0.4	0.952	0.907	0.865	0.826	0.789	0.755	0.723	0.694	0.666	0.640	0.616
0.5	0.952	0.907	0.866	0.827	0.792	0.758	0.728	0.699	0.673	0.648	0.625
0.6	0.952	0.908	0.867	0.829	0.794	0.762	0.732	0.705	0.679	0.656	0.634
0.7	0.952	0.908	0.868	0.830	0.796	0.765	0.736	0.710	0.685	0.663	0.643
0.8	0.952	0.908	0.869	0.832	0.799	0.768	0.740	0.715	0.692	0.670	0.651
0.9	0.952	0.909	0.869	0.834	0.801	0.771	0.744	0.720	0.698	0.677	0.659
1.0	0.952	0.909	0.870	0.835	0.803	0.774	0.748	0.725	0.703	0.684	0.666
1.1	0.953	0.910	0.871	0.837	0.806	0.777	0.752	0.729	0.709	0.690	0.674
1.2	0.953	0.910	0.872	0.838	0.808	0.780	0.756	0.734	0.714	0.697	0.681
1.3	0.953	0.911	0.873	0.840	0.810	0.783	0.760	0.738	0.720	0.703	0.688
1.4	0.953	0.911	0.874	0.841	0.812	0.786	0.763	0.743	0.725	0.709	0.695
1.5	0.953	0.912	0.875	0.843	0.814	0.789	0.767	0.747	0.730	0.715	0.701
1.6	0.953	0.912	0.876	0.844	0.816	0.792	0.770	0.751	0.735	0.720	0.707
1.7	0.953	0.912	0.877	0.845	0.818	0.794	0.774	0.755	0.740	0.726	0.714
1.8	0.953	0.913	0.878	0.847	0.820	0.797	0.777	0.759	0.744	0.731	0.719
1.9	0.953	0.913	0.878	0.848	0.822	0.800	0.780	0.763	0.749	0.736	0.725
2.0	0.954	0.914	0.879	0.850	0.824	0.802	0.783	0.767	0.753	0.741	0.731
2.1	0.954	0.914	0.880	0.851	0.826	0.805	0.786	0.771	0.757	0.746	0.736
2.2	0.954	0.914	0.881	0.852	0.828	0.807	0.789	0.774	0.762	0.751	0.741
2.3	0.954	0.915	0.882	0.854	0.830	0.810	0.792	0.778	0.766	0.755	0.746
2.4	0.954	0.915	0.882	0.855	0.832	0.812	0.795	0.781	0.770	0.760	0.751
2.5	0.954	0.916	0.883	0.856	0.833	0.814	0.798	0.785	0.773	0.764	0.756
2.6	0.954	0.916	0.884	0.857	0.835	0.817	0.801	0.788	0.777	0.768	0.761
2.7	0.954	0.916	0.885	0.859	0.837	0.819	0.804	0.791	0.781	0.772	0.765
2.8	0.954	0.917	0.886	0.860	0.839	0.821	0.806	0.794	0.784	0.776	0.769
2.9	0.955	0.917	0.886	0.861	0.840	0.823	0.809	0.797	0.788	0.780	0.774
3.0	0.955	0.918	0.887	0.862	0.842	0.825	0.812	0.800	0.791	0.784	0.778

Table B.1 – (continued)

β \ γ	0.60	0.70	0.80	0.90	1.00	1.10	1.20	1.30	1.40	1.50	1.60
0.1	0.561	0.512	0.468	0.429	0.394	0.362	0.334	0.308	0.286	0.265	0.247
0.2	0.572	0.526	0.486	0.450	0.418	0.389	0.364	0.342	0.322	0.304	0.289
0.3	0.583	0.540	0.503	0.470	0.440	0.415	0.392	0.373	0.355	0.340	0.327
0.4	0.594	0.554	0.519	0.488	0.462	0.439	0.419	0.401	0.386	0.373	0.362
0.5	0.604	0.567	0.534	0.506	0.482	0.461	0.444	0.428	0.415	0.404	0.394
0.6	0.614	0.579	0.549	0.523	0.501	0.483	0.467	0.453	0.442	0.432	0.423
0.7	0.624	0.591	0.563	0.539	0.519	0.502	0.488	0.467	0.466	0.458	0.451
0.8	0.633	0.602	0.576	0.554	0.536	0.521	0.509	0.498	0.489	0.482	0.476
0.9	0.642	0.613	0.589	0.569	0.552	0.539	0.528	0.518	0.510	0.504	0.499
1.0	0.651	0.623	0.601	0.583	0.568	0.555	0.545	0.537	0.530	0.525	0.520
1.1	0.659	0.633	0.613	0.596	0.582	0.571	0.562	0.555	0.549	0.544	0.540
1.2	0.667	0.643	0.624	0.608	0.596	0.586	0.578	0.571	0.566	0.562	0.559
1.3	0.675	0.652	0.634	0.620	0.609	0.600	0.593	0.587	0.583	0.579	0.576
1.4	0.682	0.661	0.644	0.631	0.621	0.613	0.607	0.602	0.598	0.595	0.592
1.5	0.689	0.670	0.654	0.642	0.633	0.626	0.620	0.616	0.612	0.609	0.607
1.6	0.696	0.678	0.663	0.652	0.644	0.637	0.632	0.628	0.625	0.623	0.621
1.7	0.703	0.686	0.672	0.662	0.655	0.649	0.644	0.641	0.638	0.636	0.635
1.8	0.709	0.693	0.681	0.672	0.665	0.659	0.655	0.652	0.650	0.648	0.647
1.9	0.716	0.700	0.689	0.681	0.674	0.669	0.666	0.663	0.661	0.660	0.659
2.0	0.722	0.707	0.697	0.689	0.683	0.679	0.676	0.673	0.672	0.670	0.669
2.1	0.728	0.714	0.704	0.697	0.692	0.688	0.685	0.683	0.682	0.681	0.680
2.2	0.733	0.721	0.712	0.705	0.700	0.697	0.694	0.692	0.691	0.690	0.689
2.3	0.739	0.727	0.719	0.713	0.708	0.705	0.703	0.701	0.700	0.699	0.699
2.4	0.744	0.733	0.725	0.720	0.716	0.713	0.711	0.709	0.708	0.708	0.707
2.5	0.749	0.739	0.732	0.727	0.723	0.720	0.719	0.717	0.716	0.716	0.715
2.6	0.754	0.745	0.738	0.733	0.730	0.728	0.726	0.725	0.724	0.723	0.723
2.7	0.759	0.750	0.744	0.739	0.736	0.734	0.733	0.732	0.731	0.731	0.730
2.8	0.764	0.755	0.749	0.745	0.743	0.741	0.740	0.739	0.738	0.738	0.737
2.9	0.768	0.760	0.755	0.751	0.749	0.747	0.746	0.745	0.745	0.744	0.744
3.0	0.773	0.765	0.760	0.757	0.755	0.753	0.752	0.751	0.751	0.751	0.750

Table B.1 – (continued)

β	γ 1.7	1.9	2.1	2.3	2.5	2.8	3.1	3.4	3.7	4.0
0.1	0.231	0.203	0.181	0.163	0.149	0.133	0.121	0.113	0.106	0.102
0.2	0.275	0.252	0.234	0.219	0.208	0.196	0.187	0.181	0.176	0.174
0.3	0.315	0.296	0.281	0.269	0.261	0.251	0.244	0.240	0.237	0.235
0.4	0.352	0.336	0.323	0.314	0.307	0.300	0.295	0.292	0.290	0.288
0.5	0.385	0.372	0.362	0.354	0.349	0.343	0.340	0.337	0.336	0.335
0.6	0.416	0.405	0.397	0.391	0.386	0.382	0.379	0.378	0.377	0.376
0.7	0.444	0.435	0.428	0.424	0.420	0.417	0.415	0.414	0.413	0.412
0.8	0.470	0.463	0.457	0.453	0.451	0.448	0.447	0.446	0.445	0.445
0.9	0.495	0.488	0.483	0.480	0.478	0.476	0.475	0.475	0.474	0.474
1.0	0.517	0.511	0.507	0.505	0.503	0.502	0.501	0.501	0.500	0.500
1.1	0.537	0.533	0.530	0.528	0.526	0.525	0.525	0.524	0.524	0.524
1.2	0.556	0.552	0.550	0.548	0.547	0.546	0.546	0.546	0.546	0.546
1.3	0.574	0.571	0.569	0.567	0.567	0.566	0.566	0.565	0.565	0.565
1.4	0.590	0.588	0.586	0.585	0.584	0.584	0.584	0.583	0.583	0.583
1.5	0.606	0.603	0.602	0.601	0.601	0.600	0.600	0.600	0.600	0.600
1.6	0.620	0.618	0.617	0.616	0.616	0.616	0.616	0.615	0.615	0.615
1.7	0.633	0.632	0.631	–	–	–	0.630	–	–	–
1.8	0.646	0.645	0.644	–	–	–	0.643	–	–	–
1.9	0.658	0.657	0.656	0.656	–	–	0.655	–	–	–
2.0	0.669	0.668	–	–	–	–	0.667	–	–	–
2.1	0.679	–	0.678	–	–	–	–	0.677	–	–
2.2	0.689	–	–	–	–	0.688	–	–	–	–
2.3	0.698	0.698	–	–	–	0.697	–	–	–	–
2.4	0.707	–	–	–	–	0.706	–	–	–	–
2.5	0.715	0.715	–	–	–	0.714	–	–	–	–
2.6	0.723	0.723	–	–	–	0.722	–	–	–	–
2.7	–	–	–	–	–	0.730	–	–	–	–
2.8	–	–	–	–	–	0.737	–	–	–	–
2.9	–	–	–	–	–	0.744	–	–	–	–
3.0	–	–	–	–	–	0.750	–	–	–	–

Table B.2 – Factor ψ for fluids in counter flow

β \ γ	0.05	0.10	0.15	0.20	0.25	0.30	0.35	0.40	0.45	0.50	0.55
0.1	0.951	0.905	0.862	0.820	0.781	0.744	0.709	0.675	0.643	0.613	0.584
0.2	0.951	0.906	0.863	0.822	0.783	0.747	0.712	0.680	0.649	0.619	0.591
0.3	0.952	0.906	0.863	0.823	0.785	0.750	0.716	0.684	0.654	0.626	0.598
0.4	0.952	0.907	0.864	0.825	0.788	0.753	0.720	0.689	0.659	0.632	0.605
0.5	0.952	0.907	0.865	0.826	0.790	0.755	0.723	0.693	0.665	0.638	0.612
0.6	0.952	0.907	0.866	0.828	0.792	0.758	0.727	0.697	0.670	0.644	0.619
0.7	0.952	0.908	0.867	0.829	0.794	0.761	0.730	0.702	0.675	0.650	0.626
0.8	0.952	0.908	0.868	0.831	0.796	0.764	0.734	0.706	0.680	0.655	0.632
0.9	0.952	0.909	0.869	0.832	0.798	0.767	0.737	0.710	0.685	0.661	0.639
1.0	0.952	0.909	0.870	0.833	0.800	0.769	0.741	0.714	0.690	0.667	0.645
1.1	0.952	0.910	0.870	0.835	0.802	0.772	0.744	0.718	0.694	0.672	0.651
1.2	0.953	0.910	0.871	0.836	0.804	0.774	0.747	0.722	0.699	0.678	0.658
1.3	0.953	0.910	0.872	0.837	0.806	0.777	0.751	0.726	0.704	0.683	0.664
1.4	0.953	0.911	0.873	0.839	0.808	0.780	0.754	0.730	0.708	0.688	0.669
1.5	0.953	0.911	0.874	0.840	0.810	0.782	0.757	0.734	0.713	0.693	0.675
1.6	0.953	0.912	0.875	0.841	0.812	0.785	0.760	0.738	0.717	0.698	0.681
1.7	0.953	0.912	0.875	0.843	0.813	0.787	0.763	0.741	0.721	0.703	0.687
1.8	0.953	0.912	0.876	0.844	0.815	0.789	0.766	0.745	0.726	0.708	0.692
1.9	0.953	0.913	0.877	0.845	0.817	0.792	0.769	0.749	0.730	0.713	0.697
2.0	0.953	0.913	0.878	0.847	0.819	0.794	0.772	0.752	0.734	0.718	0.703
2.1	0.954	0.913	0.879	0.848	0.821	0.796	0.775	0.756	0.738	0.722	0.708
2.2	0.954	0.914	0.879	0.849	0.822	0.799	0.778	0.759	0.742	0.727	0.713
2.3	0.954	0.914	0.880	0.850	0.824	0.801	0.781	0.762	0.746	0.731	0.718
2.4	0.954	0.915	0.881	0.851	0.826	0.803	0.783	0.766	0.750	0.736	0.723
2.5	0.954	0.915	0.882	0.853	0.827	0.805	0.786	0.769	0.753	0.740	0.728
2.6	0.954	0.915	0.882	0.854	0.829	0.808	0.789	0.772	0.757	0.744	0.732
2.7	0.954	0.916	0.883	0.855	0.831	0.810	0.791	0.775	0.761	0.748	0.737
2.8	0.954	0.916	0.884	0.856	0.832	0.812	0.794	0.778	0.764	0.752	0.741
2.9	0.954	0.917	0.885	0.857	0.834	0.814	0.796	0.781	0.768	0.756	0.746
3.0	0.955	0.917	0.885	0.858	0.836	0.816	0.799	0.784	0.771	0.760	0.750

Table B.2 – (continued)

β	γ 0.6	0.7	0.8	0.9	1.0	1.1	1.2	1.3	1.4	1.5	1.6
0.1	0.557	0.506	0.460	0.419	0.381	0.347	0.316	0.288	0.263	0.240	0.218
0.2	0.565	0.516	0.472	0.431	0.395	0.362	0.332	0.304	0.279	0.256	0.236
0.3	0.573	0.525	0.483	0.444	0.408	0.376	0.347	0.320	0.296	0.274	0.253
0.4	0.581	0.535	0.493	0.456	0.422	0.391	0.363	0.337	0.313	0.291	0.271
0.5	0.588	0.544	0.504	0.468	0.435	0.405	0.378	0.353	0.330	0.309	0.290
0.6	0.596	0.553	0.515	0.480	0.449	0.420	0.394	0.370	0.348	0.327	0.309
0.7	0.603	0.562	0.525	0.492	0.462	0.434	0.409	0.386	0.365	0.345	0.327
0.8	0.611	0.571	0.535	0.504	0.475	0.448	0.424	0.402	0.382	0.364	0.347
0.9	0.618	0.580	0.546	0.515	0.487	0.462	0.440	0.419	0.400	0.382	0.366
1.0	0.625	0.588	0.556	0.526	0.500	0.476	0.455	0.435	0.417	0.400	0.385
1.1	0.632	0.597	0.565	0.537	0.512	0.490	0.469	0.451	0.434	0.418	0.403
1.2	0.639	0.605	0.575	0.548	0.525	0.503	0.484	0.466	0.450	0.436	0.422
1.3	0.646	0.613	0.584	0.559	0.536	0.516	0.498	0.482	0.467	0.453	0.440
1.4	0.652	0.621	0.594	0.570	0.548	0.529	0.512	0.497	0.483	0.470	0.458
1.5	0.659	0.629	0.603	0.580	0.560	0.542	0.526	0.511	0.498	0.487	0.476
1.6	0.665	0.636	0.611	0.590	0.571	0.554	0.539	0.526	0.514	0.503	0.493
1.7	0.671	0.644	0.620	0.600	0.582	0.566	0.552	0.540	0.528	0.518	0.510
1.8	0.677	0.651	0.629	0.609	0.592	0.578	0.565	0.553	0.543	0.534	0.526
1.9	0.683	0.658	0.637	0.618	0.603	0.589	0.577	0.566	0.557	0.549	0.541
2.0	0.689	0.665	0.645	0.628	0.613	0.600	0.589	0.579	0.570	0.563	0.556
2.1	0.695	0.672	0.653	0.636	0.622	0.610	0.600	0.591	0.583	0.577	0.571
2.2	0.700	0.679	0.660	0.645	0.632	0.621	0.611	0.603	0.596	0.590	0.584
2.3	0.706	0.685	0.668	0.653	0.641	0.631	0.622	0.615	0.608	0.602	0.598
2.4	0.711	0.691	0.675	0.662	0.650	0.641	0.632	0.626	0.620	0.615	0.610
2.5	0.717	0.698	0.682	0.669	0.659	0.650	0.642	0.636	0.631	0.626	0.623
2.6	0.722	0.704	0.689	0.677	0.667	0.659	0.652	0.646	0.642	0.638	0.634
2.7	0.727	0.710	0.696	0.685	0.675	0.668	0.661	0.656	0.652	0.648	0.645
2.8	0.732	0.715	0.702	0.692	0.683	0.676	0.670	0.666	0.662	0.659	0.656
2.9	0.736	0.721	0.709	0.699	0.691	0.684	0.679	0.675	0.671	0.669	0.666
3.0	0.741	0.726	0.715	0.706	0.698	0.692	0.687	0.684	0.680	0.678	0.676

Table B.2 – (continued)

β	γ 1.7	1.9	2.1	2.3	2.5	2.8	3.1	3.4	3.7	4.0
0.1	0.199	0.166	0.138	0.115	0.096	0.073	0.056	0.042	0.032	0.025
0.2	0.216	0.183	0.155	0.131	0.111	0.087	0.068	0.053	0.042	0.033
0.3	0.234	0.201	0.173	0.149	0.128	0.103	0.083	0.067	0.054	0.043
0.4	0.253	0.220	0.192	0.168	0.147	0.121	0.100	0.082	0.068	0.056
0.5	0.272	0.240	0.212	0.188	0.167	0.141	0.119	0.101	0.085	0.073
0.6	0.291	0.260	0.233	0.210	0.189	0.162	0.140	0.121	0.105	0.092
0.7	0.311	0.281	0.255	0.232	0.212	0.186	0.164	0.145	0.129	0.114
0.8	0.331	0.302	0.277	0.255	0.236	0.210	0.189	0.170	0.154	0.140
0.9	0.351	0.323	0.300	0.279	0.260	0.236	0.216	0.198	0.183	0.169
1.0	0.370	0.345	0.323	0.303	0.286	0.263	0.244	0.227	0.213	0.200
1.1	0.390	0.366	0.346	0.327	0.311	0.291	0.273	0.258	0.244	0.233
1.2	0.410	0.387	0.368	0.352	0.337	0.318	0.302	0.288	0.277	0.266
1.3	0.429	0.408	0.391	0.376	0.362	0.346	0.331	0.319	0.309	0.300
1.4	0.448	0.429	0.413	0.399	0.388	0.373	0.360	0.350	0.341	0.334
1.5	0.466	0.449	0.435	0.423	0.412	0.399	0.388	0.380	0.372	0.366
1.6	0.484	0.469	0.456	0.445	0.436	0.424	0.415	0.408	0.402	0.398
1.7	0.502	0.488	0.476	0.467	0.459	0.449	0.441	0.435	0.431	0.427
1.8	0.518	0.506	0.496	0.487	0.481	0.472	0.466	0.461	0.458	0.455
1.9	0.535	0.524	0.515	0.507	0.502	0.495	0.490	0.486	0.483	0.481
2.0	0.550	0.540	0.533	0.526	0.521	0.516	0.512	0.508	0.506	0.505
2.1	0.565	0.557	0.550	0.544	0.540	0.536	0.532	0.530	0.528	0.527
2.2	0.580	0.572	0.566	0.562	0.558	0.554	0.552	0.550	0.548	0.548
2.3	0.594	0.587	0.582	0.578	0.575	0.572	0.570	0.568	0.567	0.567
2.4	0.607	0.601	0.596	0.593	0.591	0.588	0.587	0.585	0.585	0.584
2.5	0.619	0.614	0.610	0.608	0.606	0.604	0.602	--------	0.601	--------
2.6	0.631	0.627	0.624	0.621	0.620	0.618	0.617	--------	0.616	--------
2.7	0.643	0.639	0.636	0.634	0.633	0.632	0.631	--------	0.630	--------
2.8	0.654	0.650	0.648	0.647	0.645	0.644	0.644	--------	0.643	--------
2.9	0.664	0.661	0.659	0.658	0.657	--------	0.656	--------	0.655	0.655
3.0	0.674	0.672	0.670	0.669	0.668	--------	--------	0.667	--------	--------

B.2 Fluids in Cross Flow

Table B.3 – Factor ψ for fluids in cross flow

	γ										
β	0.05	0.10	0.15	0.20	0.25	0.30	0.35	0.40	0.45	0.50	0.55
0.1	0.951	0.905	0.862	0.820	0.781	0.744	0.709	0.676	0.644	0.614	0.586
0.2	0.951	0.906	0.863	0.822	0.784	0.747	0.713	0.681	0.650	0.621	0.594
0.3	0.952	0.906	0.864	0.824	0.786	0.751	0.717	0.686	0.656	0.628	0.602
0.4	0.952	0.907	0.865	0.825	0.788	0.754	0.721	0.691	0.662	0.635	0.610
0.5	0.952	0.907	0.865	0.827	0.791	0.757	0.725	0.696	0.668	0.642	0.618
0.6	0.952	0.908	0.866	0.828	0.793	0.760	0.729	0.701	0.674	0.649	0.625
0.7	0.952	0.908	0.867	0.830	0.795	0.763	0.733	0.705	0.679	0.655	0.633
0.8	0.952	0.908	0.868	0.831	0.797	0.766	0.737	0.710	0.685	0.662	0.640
0.9	0.952	0.909	0.869	0.833	0.799	0.769	0.740	0.714	0.690	0.668	0.647
1.0	0.952	0.909	0.870	0.834	0.801	0.771	0.744	0.719	0.695	0.674	0.654
1.1	0.953	0.910	0.871	0.836	0.804	0.774	0.748	0.723	0.700	0.680	0.660
1.2	0.953	0.910	0.872	0.837	0.806	0.777	0.751	0.727	0.705	0.685	0.667
1.3	0.953	0.910	0.873	0.838	0.808	0.780	0.754	0.731	0.710	0.691	0.673
1.4	0.953	0.911	0.873	0.840	0.810	0.782	0.758	0.735	0.715	0.696	0.679
1.5	0.953	0.911	0.874	0.841	0.812	0.785	0.761	0.739	0.720	0.702	0.685
1.6	0.953	0.912	0.875	0.843	0.814	0.788	0.764	0.743	0.724	0.707	0.691
1.7	0.953	0.912	0.876	0.844	0.815	0.790	0.767	0.747	0.728	0.712	0.697
1.8	0.953	0.913	0.877	0.845	0.817	0.793	0.770	0.751	0.733	0.717	0.702
1.9	0.953	0.913	0.878	0.847	0.819	0.795	0.774	0.754	0.737	0.722	0.708
2.0	0.954	0.913	0.878	0.848	0.821	0.797	0.777	0.758	0.741	0.726	0.713
2.1	0.954	0.914	0.879	0.849	0.823	0.800	0.779	0.761	0.745	0.731	0.718
2.2	0.954	0.914	0.880	0.850	0.825	0.802	0.782	0.765	0.749	0.735	0.723
2.3	0.954	0.915	0.881	0.852	0.826	0.804	0.785	0.768	0.753	0.740	0.728
2.4	0.954	0.915	0.882	0.853	0.828	0.807	0.788	0.771	0.757	0.744	0.733
2.5	0.954	0.915	0.882	0.854	0.830	0.809	0.791	0.775	0.761	0.748	0.737
2.6	0.954	0.916	0.883	0.855	0.832	0.811	0.793	0.778	0.764	0.752	0.742
2.7	0.954	0.916	0.884	0.857	0.833	0.813	0.796	0.781	0.768	0.756	0.746
2.8	0.954	0.916	0.885	0.858	0.835	0.815	0.799	0.784	0.771	0.760	0.750
2.9	0.955	0.917	0.885	0.859	0.837	0.817	0.801	0.787	0.775	0.764	0.755
3.0	0.955	0.917	0.886	0.860	0.838	0.820	0.804	0.790	0.778	0.768	0.759

Table B.3 – (continued)

	γ										
β	0.6	0.7	0.8	0.9	1.0	1.1	1.2	1.3	1.4	1.5	1.6
0.1	0.559	0.509	0.463	0.423	0.386	0.353	0.323	0.295	0.270	0.248	0.227
0.2	0.568	0.520	0.477	0.438	0.403	0.372	0.343	0.317	0.294	0.272	0.252
0.3	0.577	0.531	0.490	0.454	0.420	0.390	0.363	0.338	0.316	0.295	0.277
0.4	0.586	0.542	0.503	0.468	0.437	0.408	0.382	0.359	0.338	0.318	0.300
0.5	0.595	0.553	0.516	0.482	0.453	0.426	0.401	0.379	0.359	0.340	0.323
0.6	0.603	0.563	0.528	0.496	0.468	0.442	0.419	0.398	0.379	0.362	0.346
0.7	0.612	0.573	0.539	0.509	0.483	0.458	0.437	0.417	0.399	0.382	0.367
0.8	0.620	0.583	0.551	0.522	0.497	0.474	0.453	0.435	0.418	0.402	0.388
0.9	0.627	0.592	0.562	0.535	0.510	0.489	0.469	0.452	0.436	0.421	0.408
1.0	0.635	0.602	0.572	0.547	0.524	0.503	0.485	0.469	0.454	0.440	0.427
1.1	0.642	0.610	0.583	0.558	0.537	0.517	0.500	0.485	0.471	0.458	0.446
1.2	0.650	0.619	0.593	0.569	0.549	0.531	0.515	0.500	0.487	0.475	0.464
1.3	0.657	0.627	0.602	0.580	0.561	0.544	0.529	0.515	0.503	0.491	0.481
1.4	0.663	0.635	0.612	0.591	0.572	0.556	0.542	0.529	0.518	0.507	0.498
1.5	0.670	0.643	0.621	0.601	0.584	0.568	0.555	0.543	0.532	0.522	0.514
1.6	0.676	0.651	0.629	0.611	0.594	0.580	0.568	0.556	0.546	0.537	0.529
1.7	0.683	0.658	0.638	0.620	0.605	0.591	0.580	0.569	0.560	0.551	0.544
1.8	0.689	0.666	0.646	0.629	0.615	0.602	0.591	0.581	0.573	0.565	0.558
1.9	0.695	0.673	0.654	0.638	0.624	0.613	0.602	0.593	0.585	0.578	0.571
2.0	0.701	0.679	0.662	0.647	0.634	0.623	0.613	0.604	0.597	0.590	0.584
2.1	0.706	0.686	0.669	0.655	0.643	0.632	0.623	0.615	0.608	0.602	0.597
2.2	0.712	0.692	0.676	0.663	0.652	0.642	0.633	0.626	0.619	0.614	0.608
2.3	0.717	0.699	0.683	0.671	0.660	0.651	0.643	0.636	0.630	0.625	0.620
2.4	0.722	0.705	0.690	0.678	0.668	0.659	0.652	0.646	0.640	0.635	0.631
2.5	0.727	0.710	0.697	0.685	0.676	0.668	0.661	0.655	0.650	0.645	0.641
2.6	0.732	0.716	0.703	0.692	0.693	0.676	0.669	0.664	0.659	0.655	0.651
2.7	0.737	0.722	0.709	0.699	0.691	0.684	0.678	0.673	0.668	0.664	0.661
2.8	0.742	0.727	0.715	0.706	0.698	0.691	0.686	0.681	0.677	0.673	0.670
2.9	0.746	0.732	0.721	0.712	0.705	0.699	0.693	0.689	0.685	0.682	0.679
3.0	0.751	0.737	0.727	0.718	0.711	0.706	0.701	0.697	0.693	0.690	0.687

Table B.3 – (continued)

β	γ 1.7	1.9	2.1	2.3	2.5	2.8	3.1	3.4	3.7	4.0
0.1	0.209	0.176	0.150	0.127	0.108	0.085	0.068	0.054	0.043	0.035
0.2	0.234	0.203	0.177	0.154	0.135	0.111	0.093	0.077	0.065	0.055
0.3	0.260	0.229	0.203	0.181	0.162	0.138	0.119	0.103	0.089	0.078
0.4	0.284	0.255	0.230	0.208	0.190	0.166	0.146	0.129	0.115	0.103
0.5	0.308	0.280	0.256	0.235	0.217	0.194	0.174	0.157	0.143	0.130
0.6	0.331	0.304	0.282	0.262	0.244	0.222	0.202	0.186	0.171	0.159
0.7	0.353	0.328	0.307	0.288	0.271	0.249	0.231	0.215	0.201	0.188
0.8	0.375	0.351	0.331	0.313	0.297	0.277	0.259	0.244	0.230	0.218
0.9	0.396	0.373	0.354	0.338	0.323	0.303	0.287	0.272	0.259	0.248
1.0	0.416	0.395	0.377	0.361	0.348	0.329	0.314	0.300	0.288	0.278
1.1	0.435	0.416	0.399	0.384	0.372	0.355	0.340	0.328	0.317	0.307
1.2	0.454	0.436	0.420	0.407	0.395	0.379	0.366	0.354	0.344	0.335
1.3	0.472	0.455	0.441	0.428	0.417	0.403	0.391	0.380	0.371	0.363
1.4	0.489	0.473	0.460	0.449	0.439	0.425	0.414	0.405	0.396	0.389
1.5	0.505	0.491	0.479	0.468	0.459	0.447	0.437	0.429	0.421	0.415
1.6	0.521	0.508	0.497	0.487	0.479	0.468	0.459	0.451	0.445	0.439
1.7	0.537	0.524	0.514	0.505	0.498	0.488	0.480	0.473	0.467	0.462
1.8	0.551	0.540	0.531	0.523	0.516	0.507	0.500	0.494	0.488	0.484
1.9	0.565	0.555	0.546	0.539	0.533	0.525	0.518	0.513	0.509	0.505
2.0	0.579	0.569	0.561	0.555	0.549	0.542	0.536	0.532	0.528	0.524
2.1	0.592	0.583	0.576	0.570	0.565	0.558	0.553	0.549	0.546	0.543
2.2	0.604	0.596	0.589	0.584	0.580	0.574	0.570	0.566	0.563	0.561
2.3	0.616	0.608	0.603	0.598	0.594	0.589	0.585	0.582	0.579	0.577
2.4	0.627	0.620	0.615	0.611	0.607	0.603	0.599	0.597	0.594	0.593
2.5	0.638	0.632	0.627	0.623	0.620	0.616	0.613	0.611	0.609	0.607
2.6	0.648	0.643	0.638	0.635	0.632	0.629	0.626	0.624	0.622	0.621
2.7	0.658	0.653	0.649	0.646	0.643	0.640	0.638	0.636	0.635	0.634
2.8	0.667	0.663	0.659	0.657	0.654	0.652	0.650	0.648	0.647	0.646
2.9	0.677	0.672	0.669	0.667	0.665	0.663	0.661	0.660	0.659	0.658
3.0	0.685	0.682	0.679	0.677	0.675	0.673	0.671	0.670	0.669	0.669

B.3 Heat Exchangers

B.3.1 Heat Exchangers with Two Passages of Internal Fluid (Fig. 2.5)

Table B.4 – Heat exchangers – A and D type – 2 passages of external fluid – Fig. A.2 – Factor ψ

	γ										
β	0.05	0.10	0.20	0.30	0.40	0.50	0.60	0.70	0.80	0.90	1.00
0.1	0.951	0.905	0.820	0.744	0.676	0.614	0.559	0.510	0.465	0.425	0.389
0.2	0.951	0.906	0.822	0.748	0.681	0.622	0.569	0.522	0.480	0.443	0.409
0.3	0.952	0.906	0.824	0.751	0.687	0.630	0.579	0.535	0.495	0.460	0.428
0.4	0.952	0.907	0.825	0.754	0.692	0.637	0.589	0.546	0.509	0.476	0.447
0.5	0.952	0.907	0.827	0.757	0.697	0.644	0.598	0.558	0.523	0.492	0.465
0.6	0.952	0.908	0.828	0.760	0.702	0.651	0.607	0.569	0.536	0.507	0.482
0.7	0.952	0.908	0.830	0.763	0.707	0.658	0.616	0.580	0.549	0.522	0.499
0.8	0.952	0.908	0.831	0.766	0.711	0.665	0.625	0.590	0.561	0.536	0.515
0.9	0.952	0.909	0.833	0.769	0.716	0.671	0.633	0.601	0.573	0.550	0.530
1.0	0.952	0.909	0.834	0.772	0.721	0.677	0.641	0.610	0.585	0.563	0.545
1.1	0.953	0.910	0.836	0.775	0.725	0.683	0.649	0.620	0.596	0.576	0.559
1.2	0.953	0.910	0.837	0.778	0.729	0.689	0.656	0.629	0.607	0.588	0.572
1.3	0.953	0.911	0.839	0.781	0.734	0.695	0.664	0.638	0.617	0.600	0.585
1.4	0.953	0.911	0.840	0.784	0.738	0.701	0.671	0.647	0.627	0.611	0.598
1.5	0.953	0.911	0.842	0.786	0.742	0.707	0.678	0.655	0.637	0.622	0.610
1.6	0.953	0.912	0.843	0.789	0.746	0.712	0.685	0.663	0.646	0.632	0.621
1.7	0.953	0.912	0.844	0.792	0.750	0.717	0.691	0.671	0.655	0.642	0.632
1.8	0.953	0.913	0.846	0.794	0.754	0.722	0.698	0.679	0.664	0.652	0.643
1.9	0.953	0.913	0.847	0.797	0.758	0.728	0.704	0.686	0.672	0.661	0.653
2.0	0.954	0.913	0.848	0.799	0.761	0.732	0.710	0.693	0.680	0.670	0.663
2.1	0.954	0.914	0.850	0.802	0.765	0.737	0.716	0.700	0.688	0.679	0.672
2.2	0.954	0.914	0.851	0.804	0.769	0.742	0.722	0.707	0.696	0.687	0.681
2.3	0.954	0.915	0.852	0.806	0.772	0.747	0.727	0.713	0.703	0.695	0.690
2.4	0.954	0.915	0.854	0.809	0.775	0.751	0.733	0.720	0.710	0.703	0.698
2.5	0.954	0.915	0.855	0.811	0.779	0.755	0.738	0.726	0.717	0.710	0.706
2.6	0.954	0.916	0.856	0.813	0.782	0.760	0.743	0.732	0.723	0.717	0.714
2.7	0.954	0.916	0.857	0.815	0.785	0.764	0.748	0.737	0.730	0.724	0.721
2.8	0.954	0.917	0.858	0.817	0.788	0.768	0.753	0.743	0.736	0.731	0.728
2.9	0.955	0.917	0.860	0.820	0.791	0.772	0.758	0.748	0.742	0.737	0.735
3.0	0.955	0.917	0.861	0.822	0.794	0.776	0.762	0.753	0.747	0.743	0.741

Table B.4 – (continued)

β	γ 1.2	1.4	1.6	1.8	2.0	2.3	2.6	2.9	3.2	3.6	4.0
0.1	0.327	0.277	0.237	0.204	0.177	0.147	0.125	0.109	0.098	0.088	0.082
0.2	0.352	0.306	0.270	0.240	0.217	0.191	0.173	0.160	0.152	0.145	0.142
0.3	0.376	0.334	0.301	0.275	0.255	0.233	0.218	0.208	0.202	**0.198**	0.198
0.4	0.399	0.361	0.331	0.309	0.291	0.272	0.260	0.253	0.250	**0.248**	0.250
0.5	0.420	0.386	0.360	0.340	0.325	0.310	0.301	0.296	**0.294**	0.295	0.298
0.6	0.441	0.411	0.387	0.370	0.357	0.345	0.338	**0.336**	0.336	0.338	0.343
0.7	0.461	0.434	0.413	0.398	0.388	0.378	0.374	**0.373**	0.374	0.379	0.385
0.8	0.481	0.456	0.438	0.425	0.417	0.409	**0.407**	0.407	0.410	0.416	0.423
0.9	0.499	0.477	0.461	0.451	0.444	**0.438**	0.438	0.439	0.443	0.450	0.458
1.0	0.516	0.497	0.483	0.474	0.469	**0.466**	0.466	0.469	0.474	0.481	0.490
1.1	0.533	0.516	0.504	0.497	0.493	**0.491**	0.493	0.497	0.502	0.510	0.519
1.2	0.549	0.534	0.524	0.518	0.516	**0.515**	0.518	0.523	0.528	0.537	0.546
1.3	0.564	0.551	0.543	0.538	**0.537**	0.538	0.541	0.546	0.552	0.561	0.570
1.4	0.579	0.567	0.561	0.557	**0.556**	0.558	0.563	0.568	0.575	0.583	0.592
1.5	0.593	0.583	0.577	**0.575**	0.575	0.578	0.583	0.589	0.595	0.604	0.613
1.6	0.606	0.598	0.593	**0.592**	0.592	0.596	0.601	0.607	0.614	0.623	0.631
1.7	0.619	0.612	**0.608**	0.608	0.609	0.613	0.618	0.625	0.631	0.640	0.648
1.8	0.631	0.625	**0.622**	0.622	0.624	0.629	0.634	0.641	0.647	0.656	0.664
1.9	0.643	0.637	**0.636**	0.636	0.639	0.643	0.649	0.656	0.662	0.670	0.678
2.0	0.654	0.649	**0.648**	0.650	0.652	0.657	0.663	0.669	0.676	0.684	0.691
2.1	0.664	0.661	**0.660**	0.662	0.665	0.670	0.676	0.682	0.688	0.696	0.703
2.2	0.674	**0.672**	0.672	0.674	0.677	0.682	0.688	0.694	0.700	0.708	0.715
2.3	0.684	**0.682**	0.682	0.685	0.688	0.693	0.699	0.705	0.711	0.719	0.725
2.4	0.693	**0.692**	0.693	0.695	0.698	0.704	0.710	0.716	0.721	0.729	0.735
2.5	0.702	**0.701**	0.702	0.705	0.708	0.714	0.720	0.726	0.731	0.738	0.744
2.6	**0.710**	0.710	0.711	0.714	0.718	0.723	0.729	0.735	0.740	0.747	0.753
2.7	**0.718**	0.718	0.720	0.723	0.726	0.732	0.738	0.743	0.748	0.755	0.761
2.8	**0.726**	0.726	0.728	0.731	0.735	0.740	0.746	0.751	0.756	0.763	0.768
2.9	**0.733**	0.734	0.736	0.739	0.743	0.748	0.754	0.759	0.764	0.770	0.775
3.0	**0.740**	0.741	0.743	0.747	0.750	0.756	0.761	0.766	0.771	0.777	0.782

Minimum values of ψ are in bold type

Table B.5 – Heat exchangers – B and C types – 2 passages of external fluid Fig. A.3 – Factor ψ

	γ										
β	0.05	0.10	0.20	0.30	0.40	0.50	0.60	0.70	0.80	0.90	1.00
0.1	0.951	0.905	0.820	0.744	0.676	0.614	0.558	0.509	0.464	0.423	0.386
0.2	0.951	0.906	0.822	0.747	0.681	0.621	0.568	0.520	0.477	0.439	0.404
0.3	0.952	0.906	0.824	0.750	0.686	0.628	0.577	0.532	0.491	0.454	0.422
0.4	0.952	0.907	0.825	0.754	0.691	0.635	0.586	0.543	0.504	0.469	0.439
0.5	0.952	0.907	0.827	0.757	0.696	0.642	0.595	0.553	0.517	0.484	0.455
0.6	0.952	0.907	0.828	0.760	0.700	0.649	0.604	0.564	0.529	0.498	0.471
0.7	0.952	0.908	0.830	0.763	0.705	0.655	0.612	0.574	0.541	0.512	0.487
0.8	0.952	0.908	0.831	0.766	0.710	0.662	0.620	0.584	0.553	0.525	0.501
0.9	0.952	0.909	0.833	0.769	0.714	0.668	0.628	0.594	0.564	0.538	0.516
1.0	0.952	0.909	0.834	0.771	0.719	0.674	0.636	0.603	0.575	0.551	0.530
1.1	0.953	0.910	0.836	0.774	0.723	0.680	0.643	0.612	0.586	0.563	0.543
1.2	0.953	0.910	0.837	0.777	0.727	0.686	0.651	0.621	0.596	0.575	0.556
1.3	0.953	0.910	0.838	0.780	0.731	0.691	0.658	0.630	0.606	0.586	0.569
1.4	0.953	0.911	0.840	0.782	0.735	0.697	0.665	0.638	0.616	0.597	0.581
1.5	0.953	0.911	0.841	0.785	0.739	0.702	0.672	0.646	0.625	0.607	0.592
1.6	0.953	0.912	0.842	0.788	0.743	0.708	0.678	0.654	0.634	0.618	0.604
1.7	0.953	0.912	0.844	0.790	0.747	0.713	0.685	0.662	0.643	0.627	0.614
1.8	0.953	0.913	0.845	0.793	0.751	0.718	0.691	0.669	0.652	0.637	0.625
1.9	0.953	0.913	0.846	0.795	0.755	0.723	0.697	0.677	0.660	0.646	0.635
2.0	0.954	0.913	0.848	0.797	0.758	0.728	0.703	0.684	0.668	0.655	0.644
2.1	0.954	0.914	0.849	0.800	0.762	0.732	0.709	0.690	0.675	0.663	0.653
2.2	0.954	0.914	0.850	0.802	0.765	0.737	0.715	0.697	0.683	0.671	0.662
2.3	0.954	0.914	0.852	0.805	0.769	0.741	0.720	0.703	0.690	0.679	0.671
2.4	0.954	0.915	0.853	0.807	0.772	0.746	0.725	0.709	0.697	0.687	0.679
2.5	0.954	0.915	0.854	0.809	0.775	0.750	0.731	0.715	0.704	0.694	0.687
2.6	0.954	0.916	0.855	0.811	0.779	0.754	0.736	0.721	0.710	0.701	0.694
2.7	0.954	0.916	0.857	0.813	0.782	0.758	0.741	0.727	0.716	0.708	0.702
2.8	0.954	0.916	0.858	0.816	0.785	0.762	0.745	0.732	0.722	0.715	0.709
2.9	0.955	0.917	0.859	0.818	0.788	0.766	0.750	0.738	0.728	0.721	0.715
3.0	0.955	0.917	0.860	0.820	0.791	0.770	0.754	0.743	0.734	0.727	0.722

Table B.5 – (continued)

β	γ 1.2	1.4	1.6	1.8	2.0	2.3	2.6	2.9	3.2	3.6	4.0
0.1	0.323	0.272	0.230	0.195	0.167	0.134	0.108	0.089	0.075	0.060	0.050
0.2	0.345	0.297	0.257	0.225	0.198	0.167	0.143	0.124	0.110	0.095	0.084
0.3	0.366	0.321	0.284	0.254	0.229	0.200	0.177	0.159	0.145	0.130	0.119
0.4	0.386	0.344	0.310	0.282	0.259	0.232	0.210	0.193	0.180	0.165	0.154
0.5	0.406	0.367	0.335	0.310	0.288	0.263	0.243	0.227	0.214	0.200	0.189
0.6	0.425	0.389	0.360	0.336	0.316	0.293	0.274	0.260	0.247	0.234	0.223
0.7	0.444	0.410	0.383	0.361	0.343	0.322	0.305	0.291	0.279	0.267	0.256
0.8	0.462	0.430	0.406	0.385	0.369	0.349	0.334	0.321	0.310	0.298	0.288
0.9	0.479	0.450	0.427	0.409	0.394	0.375	0.361	0.349	0.339	0.328	0.318
1.0	0.495	0.469	0.448	0.431	0.417	0.400	0.387	0.376	0.367	0.357	0.347
1.1	0.511	0.487	0.467	0.452	0.439	0.424	0.412	0.402	0.393	0.383	0.375
1.2	0.526	0.504	0.486	0.472	0.460	0.447	0.435	0.426	0.418	0.409	0.401
1.3	0.541	0.520	0.504	0.491	0.481	0.468	0.458	0.449	0.441	0.433	0.425
1.4	0.555	0.536	0.521	0.509	0.500	0.488	0.478	0.470	0.464	0.455	0.448
1.5	0.569	0.551	0.537	0.526	0.518	0.507	0.498	0.491	0.484	0.477	0.470
1.6	0.581	0.565	0.553	0.543	0.535	0.525	0.517	0.510	0.504	0.497	0.491
1.7	0.594	0.579	0.567	0.558	0.551	0.542	0.534	0.528	0.522	0.516	0.510
1.8	0.606	0.592	0.581	0.573	0.566	0.558	0.551	0.545	0.540	0.533	0.528
1.9	0.617	0.604	0.594	0.587	0.580	0.573	0.566	0.561	0.556	0.550	0.545
2.0	0.628	0.616	0.607	0.600	0.594	0.587	0.581	0.576	0.571	0.566	0.562
2.1	0.638	0.627	0.619	0.612	0.607	0.600	0.595	0.590	0.586	0.581	0.577
2.2	0.648	0.638	0.630	0.624	0.619	0.613	0.608	0.604	0.600	0.595	0.591
2.3	0.658	0.648	0.641	0.636	0.631	0.625	0.620	0.616	0.613	0.608	0.605
2.4	0.667	0.658	0.652	0.646	0.642	0.637	0.632	0.628	0.625	0.621	0.618
2.5	0.675	0.667	0.661	0.657	0.653	0.648	0.643	0.640	0.637	0.633	0.630
2.6	0.684	0.676	0.671	0.666	0.663	0.658	0.654	0.651	0.648	0.644	0.642
2.7	0.692	0.685	0.680	0.676	0.672	0.668	0.664	0.661	0.658	0.655	0.652
2.8	0.699	0.693	0.688	0.684	0.681	0.677	0.673	0.671	0.668	0.665	0.663
2.9	0.707	0.701	0.696	0.693	0.690	0.686	0.683	0.680	0.677	0.675	0.673
3.0	0.714	0.708	0.704	0.701	0.698	0.694	0.691	0.689	0.686	0.684	0.682

Table B.6 – Heat exchangers – A and B type – 3 passages of external fluid – Fig. A.4 – Factor ψ (valid also for I and L type with 3 passages of external fluid)

	γ										
β	0.05	0.10	0.20	0.30	0.40	0.50	0.60	0.70	0.80	0.90	1.00
0.1	0.951	0.905	0.820	0.744	0.676	0.614	0.559	0.509	0.465	0.424	0.388
0.2	0.951	0.906	0.822	0.747	0.681	0.622	0.569	0.522	0.479	0.442	0.408
0.3	0.952	0.906	0.824	0.751	0.686	0.629	0.579	0.534	0.494	0.458	0.427
0.4	0.952	0.907	0.825	0.754	0.691	0.636	0.588	0.545	0.508	0.474	0.445
0.5	0.952	0.907	0.827	0.757	0.696	0.644	0.597	0.557	0.521	0.490	0.462
0.6	0.952	0.908	0.828	0.760	0.701	0.650	0.606	0.568	0.534	0.505	0.479
0.7	0.952	0.908	0.830	0.763	0.706	0.657	0.615	0.578	0.547	0.519	0.495
0.8	0.952	0.908	0.831	0.766	0.711	0.664	0.623	0.589	0.559	0.533	0.511
0.9	0.952	0.909	0.833	0.769	0.716	0.670	0.632	0.599	0.571	0.547	0.526
1.0	0.952	0.909	0.834	0.772	0.720	0.676	0.640	0.608	0.582	0.560	0.541
1.1	0.953	0.910	0.836	0.775	0.724	0.682	0.647	0.618	0.593	0.572	0.555
1.2	0.953	0.910	0.837	0.778	0.729	0.688	0.655	0.627	0.604	0.584	0.568
1.3	0.953	0.911	0.839	0.781	0.733	0.694	0.662	0.636	0.614	0.596	0.581
1.4	0.953	0.911	0.840	0.783	0.737	0.700	0.669	0.644	0.624	0.607	0.593
1.5	0.953	0.911	0.841	0.786	0.741	0.705	0.676	0.653	0.634	0.618	0.605
1.6	0.953	0.912	0.843	0.789	0.745	0.711	0.683	0.661	0.643	0.628	0.617
1.7	0.953	0.912	0.844	0.791	0.749	0.716	0.690	0.669	0.652	0.638	0.628
1.8	0.953	0.913	0.846	0.794	0.753	0.721	0.696	0.676	0.660	0.648	0.638
1.9	0.953	0.913	0.847	0.796	0.757	0.726	0.702	0.683	0.669	0.657	0.648
2.0	0.954	0.913	0.848	0.799	0.761	0.731	0.708	0.691	0.677	0.666	0.658
2.1	0.954	0.914	0.850	0.801	0.764	0.736	0.714	0.697	0.685	0.675	0.667
2.2	0.954	0.914	0.851	0.803	0.768	0.741	0.720	0.704	0.692	0.683	0.676
2.3	0.954	0.915	0.852	0.806	0.771	0.745	0.725	0.711	0.699	0.691	0.685
2.4	0.954	0.915	0.853	0.808	0.775	0.750	0.731	0.717	0.706	0.699	0.693
2.5	0.954	0.915	0.855	0.810	0.778	0.754	0.736	0.723	0.713	0.706	0.701
2.6	0.954	0.916	0.856	0.813	0.781	0.758	0.741	0.729	0.720	0.713	0.709
2.7	0.954	0.916	0.857	0.815	0.784	0.762	0.746	0.734	0.726	0.720	0.716
2.8	0.954	0.917	0.858	0.817	0.787	0.766	0.751	0.740	0.732	0.727	0.723
2.9	0.955	0.917	0.859	0.819	0.791	0.770	0.756	0.745	0.738	0.733	0.730
3.0	0.955	0.917	0.861	0.821	0.794	0.774	0.760	0.751	0.744	0.739	0.736

Table B.6 – (continued)

β \ γ	1.2	1.4	1.6	1.8	2.0	2.3	2.6	2.9	3.2	3.6	4.0
0.1	0.326	0.276	0.235	0.202	0.175	0.143	0.121	0.104	0.092	0.081	0.075
0.2	0.350	0.304	0.266	0.236	0.212	0.185	0.165	0.151	0.141	0.133	0.129
0.3	0.373	0.331	0.297	0.270	0.248	0.224	0.207	0.196	0.188	0.182	0.180
0.4	0.395	0.356	0.326	0.302	0.283	0.262	0.248	0.239	0.233	0.229	0.228
0.5	0.417	0.381	0.354	0.332	0.316	0.298	0.286	0.279	0.275	**0.273**	0.273
0.6	0.437	0.405	0.380	0.361	0.347	0.332	0.322	0.317	**0.314**	0.314	0.316
0.7	0.457	0.427	0.405	0.389	0.376	0.364	0.357	0.353	**0.352**	0.353	0.356
0.8	0.476	0.449	0.429	0.415	0.404	0.394	0.389	**0.386**	0.386	0.388	0.392
0.9	0.494	0.470	0.452	0.440	0.431	0.423	0.419	**0.418**	0.419	0.422	0.426
1.0	0.511	0.489	0.474	0.463	0.456	0.449	**0.447**	0.447	0.449	0.453	0.458
1.1	0.527	0.508	0.495	0.485	0.479	0.475	**0.473**	0.474	0.477	0.481	0.487
1.2	0.543	0.526	0.514	0.506	0.502	**0.498**	0.498	0.499	0.502	0.508	0.514
1.3	0.558	0.543	0.533	0.526	0.522	**0.520**	0.521	0.523	0.526	0.532	0.538
1.4	0.573	0.559	0.550	0.545	0.542	**0.541**	0.542	0.545	0.549	0.554	0.561
1.5	0.587	0.574	0.567	0.563	0.561	**0.560**	0.562	0.565	0.569	0.575	0.582
1.6	0.600	0.589	0.583	0.579	**0.578**	0.578	0.581	0.584	0.588	0.594	0.601
1.7	0.612	0.603	0.598	0.595	**0.594**	0.595	0.598	0.602	0.606	0.612	0.619
1.8	0.624	0.616	0.612	**0.610**	0.610	0.611	0.614	0.618	0.623	0.629	0.635
1.9	0.636	0.629	0.625	**0.624**	0.624	0.626	0.629	0.633	0.638	0.644	0.650
2.0	0.647	0.641	0.638	**0.637**	0.638	0.640	0.644	0.648	0.652	0.658	0.664
2.1	0.657	0.652	0.650	**0.649**	0.650	0.653	0.657	0.661	0.665	0.671	0.677
2.2	0.667	0.663	**0.661**	0.661	0.662	0.665	0.669	0.673	0.678	0.684	0.689
2.3	0.677	0.673	**0.672**	0.672	0.674	0.677	0.681	0.685	0.689	0.695	0.701
2.4	0.686	0.683	**0.682**	0.683	0.684	0.688	0.692	0.696	0.700	0.706	0.711
2.5	0.695	**0.692**	0.692	0.693	0.695	0.698	0.702	0.706	0.710	0.716	0.721
2.6	0.703	**0.701**	0.701	0.702	0.704	0.708	0.712	0.716	0.720	0.725	0.730
2.7	0.711	**0.710**	0.710	0.711	0.713	0.717	0.721	0.725	0.729	0.734	0.739
2.8	0.719	**0.718**	0.718	0.720	0.722	0.725	0.729	0.733	0.737	0.742	0.747
2.9	0.726	**0.725**	0.726	0.728	0.730	0.733	0.737	0.741	0.745	0.750	0.755
3.0	**0.733**	0.733	0.734	0.735	0.737	0.741	0.745	0.749	0.752	0.757	0.762

Minimum values of ψ are in bold type

Table B.7 – Heat exchangers – C and D type – 3 passages of external fluid – Fig. A.5 – Factor ψ

	γ										
β	0.05	0.10	0.20	0.30	0.40	0.50	0.60	0.70	0.80	0.90	1.00
0.1	0.951	0.905	0.820	0.744	0.676	0.614	0.559	0.509	0.464	0.423	0.387
0.2	0.951	0.906	0.822	0.747	0.681	0.621	0.568	0.521	0.478	0.440	0.406
0.3	0.952	0.906	0.824	0.751	0.686	0.629	0.578	0.532	0.492	0.456	0.424
0.4	0.952	0.907	0.825	0.754	0.691	0.636	0.587	0.544	0.505	0.471	0.441
0.5	0.952	0.907	0.827	0.757	0.696	0.643	0.596	0.555	0.518	0.486	0.458
0.6	0.952	0.907	0.828	0.760	0.701	0.649	0.605	0.565	0.531	0.501	0.474
0.7	0.952	0.908	0.830	0.763	0.705	0.656	0.613	0.576	0.543	0.515	0.490
0.8	0.952	0.908	0.831	0.766	0.710	0.662	0.621	0.586	0.555	0.528	0.505
0.9	0.952	0.909	0.833	0.769	0.715	0.669	0.629	0.596	0.567	0.542	0.520
1.0	0.952	0.909	0.834	0.772	0.719	0.675	0.637	0.605	0.578	0.554	0.534
1.1	0.953	0.910	0.836	0.774	0.724	0.681	0.645	0.614	0.589	0.566	0.548
1.2	0.953	0.910	0.837	0.777	0.728	0.687	0.652	0.623	0.599	0.578	0.561
1.3	0.953	0.910	0.838	0.780	0.732	0.692	0.660	0.632	0.609	0.590	0.573
1.4	0.953	0.911	0.840	0.783	0.736	0.698	0.667	0.641	0.619	0.601	0.585
1.5	0.953	0.911	0.841	0.785	0.740	0.703	0.673	0.649	0.628	0.611	0.597
1.6	0.953	0.912	0.843	0.788	0.744	0.709	0.680	0.657	0.637	0.622	0.608
1.7	0.953	0.912	0.844	0.790	0.748	0.714	0.687	0.664	0.646	0.631	0.619
1.8	0.953	0.913	0.845	0.793	0.752	0.719	0.693	0.672	0.655	0.641	0.630
1.9	0.953	0.913	0.847	0.795	0.756	0.724	0.699	0.679	0.663	0.650	0.639
2.0	0.954	0.913	0.848	0.798	0.759	0.729	0.705	0.686	0.671	0.659	0.649
2.1	0.954	0.914	0.849	0.800	0.763	0.734	0.711	0.693	0.679	0.667	0.658
2.2	0.954	0.914	0.851	0.803	0.766	0.738	0.717	0.700	0.686	0.676	0.667
2.3	0.954	0.915	0.852	0.805	0.770	0.743	0.722	0.706	0.693	0.684	0.676
2.4	0.954	0.915	0.853	0.807	0.773	0.747	0.727	0.712	0.700	0.691	0.684
2.5	0.954	0.915	0.854	0.810	0.776	0.751	0.733	0.718	0.707	0.698	0.692
2.6	0.954	0.916	0.856	0.812	0.780	0.756	0.738	0.724	0.714	0.706	0.699
2.7	0.954	0.916	0.857	0.814	0.783	0.760	0.743	0.730	0.720	0.712	0.706
2.8	0.954	0.916	0.858	0.816	0.786	0.764	0.747	0.735	0.726	0.719	0.713
2.9	0.955	0.917	0.859	0.818	0.789	0.768	0.752	0.740	0.732	0.725	0.720
3.0	0.955	0.917	0.860	0.820	0.792	0.771	0.757	0.746	0.737	0.731	0.727

Table B.7 – (continued)

	γ										
β	1.2	1.4	1.6	1.8	2.0	2.3	2.6	2.9	3.2	3.6	4.0
0.1	0.324	0.273	0.232	0.198	0.170	0.137	0.113	0.094	0.081	0.067	0.058
0.2	0.347	0.299	0.261	0.229	0.203	0.173	0.150	0.133	0.120	0.107	0.098
0.3	0.369	0.325	0.289	0.260	0.236	0.208	0.187	0.172	0.159	0.147	0.137
0.4	0.390	0.349	0.316	0.289	0.268	0.242	0.223	0.209	0.197	0.185	0.176
0.5	0.410	0.372	0.342	0.318	0.298	0.275	0.258	0.244	0.234	0.223	0.214
0.6	0.430	0.395	0.367	0.345	0.327	0.306	0.291	0.279	0.269	0.259	0.250
0.7	0.449	0.417	0.391	0.371	0.355	0.336	0.322	0.311	0.302	0.293	0.284
0.8	0.467	0.437	0.414	0.396	0.382	0.365	0.352	0.342	0.334	0.325	0.317
0.9	0.484	0.457	0.436	0.420	0.407	0.392	0.380	0.371	0.363	0.355	0.347
1.0	0.501	0.476	0.457	0.442	0.431	0.417	0.407	0.398	0.391	0.383	0.376
1.1	0.517	0.494	0.477	0.464	0.453	0.441	0.431	0.424	0.417	0.410	0.403
1.2	0.533	0.512	0.496	0.484	0.474	0.463	0.455	0.448	0.442	0.435	0.428
1.3	0.547	0.528	0.514	0.503	0.495	0.485	0.477	0.470	0.465	0.458	0.452
1.4	0.562	0.544	0.531	0.521	0.514	0.505	0.497	0.491	0.486	0.480	0.474
1.5	0.575	0.559	0.547	0.539	0.532	0.523	0.517	0.511	0.506	0.500	0.494
1.6	0.588	0.574	0.563	0.555	0.548	0.541	0.535	0.530	0.525	0.519	0.514
1.7	0.600	0.587	0.577	0.570	0.564	0.557	0.552	0.547	0.542	0.537	0.532
1.8	0.612	0.600	0.591	0.585	0.579	0.573	0.568	0.563	0.559	0.554	0.549
1.9	0.624	0.613	0.605	0.598	0.594	0.588	0.583	0.578	0.574	0.569	0.564
2.0	0.634	0.624	0.617	0.611	0.607	0.602	0.597	0.593	0.589	0.584	0.579
2.1	0.645	0.636	0.629	0.624	0.620	0.615	0.610	0.606	0.602	0.598	0.593
2.2	0.655	0.646	0.640	0.635	0.632	0.627	0.623	0.619	0.615	0.611	0.607
2.3	0.664	0.656	0.651	0.646	0.643	0.638	0.634	0.631	0.627	0.623	0.619
2.4	0.673	0.666	0.661	0.657	0.654	0.649	0.646	0.642	0.639	0.635	0.631
2.5	0.682	0.675	0.671	0.667	0.664	0.660	0.656	0.653	0.650	0.646	0.642
2.6	0.690	0.684	0.680	0.676	0.673	0.669	0.666	0.663	0.660	0.656	0.653
2.7	0.698	0.692	0.688	0.685	0.682	0.679	0.676	0.672	0.670	0.666	0.663
2.8	0.706	0.701	0.697	0.694	0.691	0.688	0.684	0.682	0.679	0.676	0.673
2.9	0.713	0.708	0.705	0.702	0.699	0.696	0.693	0.690	0.688	0.685	0.682
3.0	0.720	0.716	0.712	0.709	0.707	0.704	0.701	0.698	0.696	0.693	0.691

Table B.8 – Heat exchangers – A and D type – 4 passages of external fluid – Fig. A.6 – Factor ψ (valid also for I and N type with 4 passages of external fluid)

β \ γ	0.05	0.10	0.20	0.30	0.40	0.50	0.60	0.70	0.80	0.90	1.00
0.1	0.951	0.905	0.820	0.744	0.676	0.614	0.559	0.509	0.464	0.424	0.388
0.2	0.951	0.906	0.822	0.747	0.681	0.622	0.569	0.521	0.479	0.441	0.407
0.3	0.952	0.906	0.824	0.751	0.686	0.629	0.578	0.533	0.493	0.458	0.426
0.4	0.952	0.907	0.825	0.754	0.691	0.636	0.588	0.545	0.507	0.474	0.444
0.5	0.952	0.907	0.827	0.757	0.696	0.643	0.597	0.556	0.521	0.489	0.461
0.6	0.952	0.908	0.828	0.760	0.701	0.650	0.606	0.567	0.533	0.504	0.478
0.7	0.952	0.908	0.830	0.763	0.706	0.657	0.614	0.578	0.546	0.518	0.494
0.8	0.952	0.908	0.831	0.766	0.711	0.663	0.623	0.588	0.558	0.532	0.510
0.9	0.952	0.909	0.833	0.769	0.715	0.670	0.631	0.598	0.570	0.546	0.525
1.0	0.952	0.909	0.834	0.772	0.720	0.676	0.639	0.608	0.581	0.559	0.539
1.1	0.953	0.910	0.836	0.775	0.724	0.682	0.647	0.617	0.592	0.571	0.553
1.2	0.953	0.910	0.837	0.778	0.729	0.688	0.654	0.626	0.603	0.583	0.566
1.3	0.953	0.910	0.839	0.780	0.733	0.694	0.662	0.635	0.613	0.595	0.579
1.4	0.953	0.911	0.840	0.783	0.737	0.699	0.669	0.644	0.623	0.606	0.592
1.5	0.953	0.911	0.841	0.786	0.741	0.705	0.676	0.652	0.632	0.616	0.603
1.6	0.953	0.912	0.843	0.788	0.745	0.710	0.682	0.660	0.642	0.627	0.615
1.7	0.953	0.912	0.844	0.791	0.749	0.716	0.689	0.668	0.651	0.637	0.626
1.8	0.953	0.913	0.846	0.794	0.753	0.721	0.695	0.675	0.659	0.646	0.636
1.9	0.954	0.913	0.847	0.796	0.757	0.726	0.702	0.683	0.668	0.656	0.646
2.0	0.954	0.913	0.848	0.798	0.760	0.731	0.708	0.690	0.676	0.665	0.656
2.1	0.954	0.914	0.849	0.801	0.764	0.735	0.713	0.696	0.683	0.673	0.665
2.2	0.954	0.914	0.851	0.803	0.767	0.740	0.719	0.703	0.691	0.681	0.674
2.3	0.954	0.915	0.852	0.806	0.771	0.745	0.725	0.710	0.698	0.689	0.683
2.4	0.954	0.915	0.853	0.808	0.774	0.749	0.730	0.716	0.705	0.697	0.691
2.5	0.954	0.915	0.855	0.810	0.778	0.753	0.735	0.722	0.712	0.704	0.699
2.6	0.954	0.916	0.856	0.812	0.781	0.758	0.740	0.728	0.718	0.712	0.707
2.7	0.954	0.916	0.857	0.815	0.784	0.762	0.745	0.733	0.725	0.718	0.714
2.8	0.954	0.916	0.858	0.817	0.787	0.766	0.750	0.739	0.731	0.725	0.721
2.9	0.955	0.917	0.859	0.819	0.790	0.770	0.755	0.744	0.737	0.731	0.728
3.0	0.955	0.917	0.861	0.821	0.793	0.773	0.759	0.749	0.742	0,737	0.734

Table B.8 – (continued)

β \ γ	1.2	1.4	1.6	1.8	2.0	2.3	2.6	2.9	3.2	3.6	4.0
0.1	0.326	0.275	0.234	0.201	0.174	0.142	0.119	0.102	0.090	0.078	0.071
0.2	0.349	0.303	0.265	0.235	0.210	0.182	0.162	0.147	0.137	0.128	0.122
0.3	0.372	0.329	0.295	0.268	0.246	0.221	0.203	0.191	0.182	0.175	0.171
0.4	0.394	0.355	0.324	0.299	0.280	0.258	0.243	0.232	0.226	0.220	0.218
0.5	0.415	0.379	0.351	0.329	0.312	0.293	0.280	0.272	0.267	0.263	0.262
0.6	0.436	0.403	0.377	0.358	0.343	0.326	0.316	0.309	0.305	**0.303**	0.303
0.7	0.455	0.425	0.402	0.385	0.372	0.358	0.349	0.344	0.342	**0.341**	0.342
0.8	0.474	0.447	0.426	0.411	0.400	0.388	0.381	0.377	**0.376**	0.376	0.378
0.9	0.492	0.467	0.449	0.436	0.426	0.416	0.411	0.408	**0.407**	0.409	0.411
1.0	0.509	0.487	0.470	0.459	0.451	0.443	0.439	**0.437**	0.437	0.439	0.442
1.1	0.525	0.505	0.491	0.481	0.474	0.468	0.465	**0.464**	0.465	0.467	0.471
1.2	0.541	0.523	0.510	0.502	0.496	0.491	**0.489**	0.489	0.490	0.493	0.497
1.3	0.556	0.540	0.529	0.521	0.517	0.513	**0.512**	0.512	0.514	0.518	0.522
1.4	0.570	0.556	0.546	0.540	0.536	**0.533**	0.533	0.534	0.536	0.540	0.544
1.5	0.584	0.571	0.563	0.558	0.555	**0.553**	0.553	0.554	0.557	0.561	0.565
1.6	0.597	0.586	0.579	0.574	0.572	**0.571**	0.571	0.573	0.576	0.580	0.585
1.7	0.610	0.600	0.593	0.590	**0.588**	0.588	0.589	0.591	0.594	0.598	0.602
1.8	0.622	0.613	0.607	0.605	**0.603**	0.603	0.605	0.607	0.610	0.614	0.619
1.9	0.633	0.625	0.621	0.619	**0.618**	0.618	0.620	0.623	0.626	0.630	0.635
2.0	0.644	0.637	0.633	0.632	**0.631**	0.632	0.634	0.637	0.640	0.644	0.649
2.1	0.655	0.649	0.645	**0.644**	0.644	0.645	0.647	0.650	0.653	0.658	0.662
2.2	0.665	0.659	0.657	**0.656**	0.656	0.658	0.660	0.663	0.666	0.670	0.675
2.3	0.674	0.670	0.668	**0.667**	0.667	0.669	0.672	0.674	0.678	0.682	0.686
2.4	0.683	0.679	0.678	**0.677**	0.678	0.680	0.683	0.685	0.689	0.693	0.697
2.5	0.692	0.689	**0.687**	0.687	0.688	0.690	0.693	0.696	0.699	0.703	0.707
2.6	0.700	0.698	**0.697**	0.697	0.698	0.700	0.703	0.706	0.709	0.713	0.717
2.7	0.708	0.706	**0.705**	0.706	0.707	0.709	0.712	0.715	0.718	0.722	0.726
2.8	0.716	**0.714**	0.714	0.714	0.715	0.718	0.720	0.723	0.726	0.730	0.734
2.9	0.723	**0.722**	0.722	0.722	0.724	0.726	0.729	0.732	0.735	0.738	0.742
3.0	0.730	**0.729**	0.729	0.730	0.731	0.734	0.736	0.739	0.742	0.746	0.750

Minimum values of ψ are in bold type

Table B.9 – Heat exchangers – B and C type – 4 passages of external fluid – Fig. A.7 – Factor ψ

	γ										
β	0.05	0.10	0.20	0.30	0.40	0.50	0.60	0.70	0.80	0.90	1.00
0.1	0.951	0.905	0.820	0.744	0.676	0.614	0.559	0.509	0.464	0.424	0.387
0.2	0.951	0.906	0.822	0.747	0.681	0.622	0.568	0.521	0.478	0.440	0.406
0.3	0.952	0.906	0.824	0.751	0.686	0.629	0.578	0.533	0.492	0.456	0.424
0.4	0.952	0.907	0.825	0.754	0.691	0.636	0.587	0.544	0.506	0.472	0.442
0.5	0.952	0.907	0.827	0.757	0.696	0.643	0.596	0.555	0.519	0.487	0.459
0.6	0.952	0.907	0.828	0.760	0.701	0.650	0.605	0.566	0.532	0.502	0.475
0.7	0.952	0.908	0.830	0.763	0.706	0.656	0.613	0.576	0.544	0.516	0.491
0.8	0.952	0.908	0.831	0.766	0.710	0.663	0.622	0.586	0.556	0.529	0.506
0.9	0.952	0.909	0.833	0.769	0.715	0.669	0.630	0.596	0.568	0.543	0.521
1.0	0.952	0.909	0.834	0.772	0.719	0.675	0.638	0.606	0.579	0.555	0.535
1.1	0.953	0.910	0.836	0.775	0.724	0.681	0.645	0.615	0.590	0.568	0.549
1.2	0.953	0.910	0.837	0.777	0.728	0.687	0.653	0.624	0.600	0.580	0.562
1.3	0.953	0.910	0.839	0.780	0.732	0.693	0.660	0.633	0.610	0.591	0.575
1.4	0.953	0.911	0.840	0.783	0.736	0.698	0.667	0.641	0.620	0.602	0.587
1.5	0.953	0.911	0.841	0.785	0.740	0.704	0.674	0.650	0.629	0.613	0.599
1.6	0.953	0.912	0.843	0.788	0.744	0.709	0.681	0.658	0.639	0.623	0.610
1.7	0.953	0.912	0.844	0.791	0.748	0.714	0.687	0.665	0.647	0.633	0.621
1.8	0.953	0.913	0.845	0.793	0.752	0.720	0.694	0.673	0.656	0.642	0.631
1.9	0.953	0.913	0.847	0.796	0.756	0.725	0.700	0.680	0.664	0.652	0.641
2.0	0.954	0.913	0.848	0.798	0.759	0.729	0.706	0.687	0.672	0.660	0.651
2.1	0.954	0.914	0.849	0.800	0.763	0.734	0.712	0.694	0.680	0.669	0.660
2.2	0.954	0.914	0.851	0.803	0.767	0.739	0.717	0.701	0.687	0.677	0.669
2.3	0.954	0.915	0.852	0.805	0.770	0.743	0.723	0.707	0.695	0.685	0.678
2.4	0.954	0.915	0.853	0.807	0.773	0.748	0.728	0.713	0.702	0.693	0.686
2.5	0.954	0.915	0.854	0.810	0.777	0.752	0.733	0.719	0.708	0.700	0.694
2.6	0.954	0.916	0.856	0.812	0.780	0.756	0.738	0.725	0.715	0.707	0.701
2.7	0.954	0.916	0.857	0.814	0.783	0.760	0.743	0.731	0.721	0.714	0.708
2.8	0.954	0.916	0.858	0.816	0.786	0.764	0.748	0.736	0.727	0.721	0.715
2.9	0.955	0.917	0.859	0.818	0.789	0.768	0.753	0.742	0.733	0.727	0.722
3.0	0.955	0.917	0.860	0.821	0.792	0.772	0.757	0.747	0.739	0.733	0.729

Table B.9 – (continued)

β	γ 1.2	1.4	1.6	1.8	2.0	2.3	2.6	2.9	3.2	3.6	4.0
0.1	0.325	0.274	0.232	0.198	0.171	0.138	0.114	0.096	0.083	0.070	0.061
0.2	0.348	0.300	0.262	0.231	0.205	0.176	0.154	0.137	0.125	0.112	0.104
0.3	0.370	0.326	0.291	0.262	0.239	0.212	0.192	0.177	0.165	0.154	0.146
0.4	0.391	0.351	0.318	0.292	0.271	0.247	0.228	0.215	0.205	0.194	0.186
0.5	0.412	0.374	0.345	0.321	0.302	0.280	0.264	0.252	0.242	0.233	0.225
0.6	0.431	0.397	0.370	0.349	0.331	0.312	0.297	0.286	0.278	0.269	0.263
0.7	0.450	0.419	0.394	0.375	0.360	0.342	0.329	0.320	0.312	0.304	0.298
0.8	0.469	0.440	0.418	0.400	0.386	0.371	0.359	0.351	0.344	0.337	0.331
0.9	0.486	0.460	0.440	0.424	0.412	0.398	0.388	0.380	0.374	0.368	0.362
1.0	0.503	0.479	0.461	0.447	0.436	0.424	0.415	0.408	0.402	0.396	0.391
1.1	0.519	0.497	0.481	0.468	0.459	0.448	0.440	0.434	0.429	0.423	0.418
1.2	0.535	0.515	0.500	0.489	0.480	0.470	0.463	0.458	0.453	0.448	0.443
1.3	0.550	0.531	0.518	0.508	0.500	0.492	0.485	0.480	0.476	0.471	0.466
1.4	0.564	0.547	0.535	0.526	0.519	0.512	0.506	0.501	0.497	0.492	0.488
1.5	0.578	0.562	0.552	0.543	0.537	0.530	0.525	0.521	0.517	0.513	0.508
1.6	0.591	0.577	0.567	0.560	0.554	0.548	0.543	0.539	0.536	0.531	0.527
1.7	0.603	0.590	0.582	0.575	0.570	0.565	0.560	0.557	0.553	0.549	0.545
1.8	0.615	0.603	0.596	0.590	0.585	0.580	0.576	0.573	0.569	0.565	0.561
1.9	0.626	0.616	0.609	0.603	0.599	0.595	0.591	0.588	0.585	0.581	0.577
2.0	0.637	0.628	0.621	0.616	0.613	0.608	0.605	0.602	0.599	0.595	0.591
2.1	0.647	0.639	0.633	0.629	0.625	0.621	0.618	0.615	0.612	0.608	0.605
2.2	0.657	0.650	0.644	0.640	0.637	0.634	0.630	0.627	0.625	0.621	0.617
2.3	0.667	0.660	0.655	0.651	0.648	0.645	0.642	0.639	0.636	0.633	0.630
2.4	0.676	0.669	0.665	0.662	0.659	0.656	0.653	0.650	0.648	0.644	0.641
2.5	0.684	0.679	0.675	0.672	0.669	0.666	0.663	0.661	0.658	0.655	0.652
2.6	0.693	0.687	0.684	0.681	0.679	0.676	0.673	0.671	0.668	0.665	0.662
2.7	0.701	0.696	0.692	0.690	0.688	0.685	0.682	0.680	0.677	0.674	0.671
2.8	0.708	0.704	0.701	0.698	0.696	0.694	0.691	0.689	0.686	0.683	0.681
2.9	0.716	0.711	0.709	0.706	0.704	0.702	0.699	0.697	0.695	0.692	0.689
3.0	0.723	0.719	0.716	0.714	0.712	0.710	0.707	0.705	0.703	0.700	0.698

Table B.10 – Heat exchangers – A and B type – 5 passages of external fluid – Fig. A.8 – Factor ψ (valid also for I and L type with 5 passages of external fluid)

	γ										
β	0.05	0.10	0.20	0.30	0.40	0.50	0.60	0.70	0.80	0.90	1.00
0.1	0.951	0.905	0.820	0.744	0.676	0.614	0.559	0.509	0.464	0.424	0.388
0.2	0.951	0.906	0.822	0.747	0.681	0.622	0.569	0.521	0.479	0.441	0.407
0.3	0.952	0.906	0.824	0.751	0.686	0.629	0.578	0.533	0.493	0.457	0.426
0.4	0.952	0.907	0.825	0.754	0.691	0.636	0.588	0.545	0.507	0.473	0.444
0.5	0.952	0.907	0.827	0.757	0.696	0.643	0.597	0.556	0.520	0.489	0.461
0.6	0.952	0.908	0.828	0.760	0.701	0.650	0.606	0.567	0.533	0.504	0.478
0.7	0.952	0.908	0.830	0.763	0.706	0.657	0.614	0.578	0.546	0.518	0.494
0.8	0,952	0.908	0.831	0.766	0.711	0.663	0.623	0.588	0.558	0.532	0.509
0.9	0.952	0.909	0.833	0.769	0.715	0.670	0.631	0.598	0.569	0.545	0.524
1.0	0.952	0.909	0.834	0.772	0.720	0.676	0.639	0.607	0.581	0.558	0.539
1.1	0.953	0.910	0.836	0.775	0.724	0.682	0.647	0.617	0.592	0.570	0.552
1.2	0.953	0.910	0.837	0.778	0.729	0.688	0.654	0.626	0.602	0.582	0.566
1.3	0.953	0.910	0.839	0.780	0.733	0.694	0.661	0.635	0.612	0.594	0.579
1.4	0.953	0.911	0.840	0.783	0.737	0.699	0.669	0.643	0.622	0.605	0.591
1.5	0.953	0.911	0.841	0.786	0.741	0.705	0.675	0.651	0.632	0.616	0.603
1.6	0.953	0.912	0.843	0.788	0.745	0.710	0.682	0.659	0.641	0.626	0.614
1.7	0.953	0.912	0.844	0.791	0.749	0.715	0.689	0.667	0.650	0.636	0.625
1.8	0.953	0.913	0.845	0.793	0.753	0.721	0.695	0.675	0.659	0.646	0.635
1.9	0.953	0.913	0.847	0.796	0.756	0.726	0.701	0.682	0.667	0.655	0.645
2.0	0.954	0.913	0.848	0.798	0.760	0.730	0.707	0.689	0.675	0.664	0.655
2.1	0.954	0.914	0.849	0.801	0.764	0.735	0.713	0.696	0.683	0.672	0.664
2.2	0.954	0.914	0.851	0.803	0.767	0.740	0.719	0.703	0.690	0.681	0.673
2.3	0.954	0.915	0.852	0.806	0.771	0.744	0.724	0.709	0.697	0.689	0.682
2.4	0.954	0.915	0.853	0.808	0.774	0.749	0.730	0.715	0.704	0.696	0.690
2.5	0.954	0.915	0.855	0.810	0.777	0.753	0.735	0.721	0.711	0.704	0.698
2.6	0.954	0.916	0.856	0.812	0.781	0.757	0.740	0.727	0.718	0.711	0.706
2.7	0.954	0.916	0.857	0.815	0.784	0.761	0.745	0.733	0.724	0.718	0.713
2.8	0.954	0.916	0.858	0.817	0.787	0.765	0.750	0.738	0.730	0.724	0.720
2.9	0.955	0.917	0.859	0.819	0.790	0.769	0.755	0.744	0.736	0.731	0.727
3.0	0.955	0.917	0.861	0.821	0.793	0.773	0.759	0.749	0.742	0.737	0.733

Table B.10 – (continued)

β	γ 1.2	1.4	1.6	1.8	2.0	2.3	2.6	2.9	3.2	3.6	4.0
0.1	0.326	0.275	0.234	0.200	0.173	0.141	0.118	0.101	0.089	0.077	0.070
0.2	0.349	0.302	0.265	0.234	0.209	0.181	0.161	0.146	0.135	0.125	0.119
0.3	0.372	0.329	0.294	0.267	0.245	0.219	0.201	0.188	0.179	0.172	0.167
0.4	0.394	0.354	0.323	0.298	0.278	0.256	0.240	0.229	0.222	0.216	0.213
0.5	0.415	0.378	0.350	0.328	0.310	0.291	0.277	0.268	0.262	0.258	0.256
0.6	0.435	0.402	0.376	0.356	0.341	0.324	0.313	0.305	0.301	0.297	0.296
0.7	0.454	0.424	0.401	0.383	0.370	0.355	0.346	0.340	0.337	**0.334**	0.334
0.8	0.473	0.445	0.425	0.409	0.397	0.385	0.377	0.373	0.370	**0.369**	0.370
0.9	0.491	0.466	0.447	0.434	0.423	0.413	0.407	0.403	**0.402**	0.402	0.403
1.0	0.508	0.485	0.469	0.457	0.448	0.439	0.434	0.432	**0.431**	0.432	0.434
1.1	0.524	0.504	0.489	0.479	0.471	0.464	0.460	**0.459**	0.459	0.460	0.462
1.2	0.540	0.522	0.509	0.499	0.493	0.487	0.485	**0.484**	0.484	0.486	0.488
1.3	0.555	0.538	0.527	0.519	0.514	0.509	**0.507**	0.507	0.508	0.510	0.513
1.4	0.569	0.554	0.544	0.538	0.533	0.530	**0.528**	0.529	0.530	0.532	0.535
1.5	0.583	0.570	0.561	0.555	0.552	0.549	**0.548**	0.549	0.550	0.553	0.556
1.6	0.596	0.584	0.577	0.572	0.569	**0.567**	0.567	0.568	0.569	0.572	0.575
1.7	0.609	0.598	0.591	0.587	0.585	**0.584**	0.584	0.585	0.587	0.590	0.593
1.8	0.621	0.611	0.605	0.602	0.600	**0.599**	0.600	0.601	0.603	0.606	0.610
1.9	0.632	0.624	0.619	0.616	0.615	**0.614**	0.615	0.617	0.619	0.622	0.625
2.0	0.643	0.636	0.631	0.629	**0.628**	0.628	0.629	0.631	0.633	0.636	0.640
2.1	0.653	0.647	0.643	**0.641**	0.641	0.641	0.642	0.644	0.647	0.650	0.653
2.2	0.663	0.658	0.655	**0.653**	0.653	0.653	0.655	0.657	0.659	0.662	0.666
2.3	0.673	0.668	0.665	**0.664**	0.664	0.665	0.667	0.669	0.671	0.674	0.678
2.4	0.682	0.678	**0.675**	0.675	0.675	0.676	0.678	0.680	0.682	0.685	0.689
2.5	0.691	0.687	**0.685**	0.685	0.685	0.686	0.688	0.690	0.692	0.696	0.699
2.6	0.699	0.696	**0.694**	0.694	0.694	0.696	0.698	0.700	0.702	0.705	0.709
2.7	0.707	0.704	**0.703**	0.703	0.704	0.705	0.707	0.709	0.711	0.714	0.718
2.8	0.715	0.712	**0.711**	0.711	0.712	0.714	0.716	0.718	0.720	0.723	0.726
2.9	0.722	0.720	**0.719**	0.720	0.720	0.722	0.724	0.726	0.728	0.731	0.734
3.0	0.729	**0.727**	0.727	0.727	0.728	0.730	0.732	0.734	0.736	0.739	0.742

Minimum values of ψ are in bold type

Table B.11 – Heat exchangers – C and D type – 5 passages of external fluid – Fig. A.9 – Factor ψ

β \ γ	0.05	0.10	0.20	0.30	0.40	0.50	0.60	0.70	0.80	0.90	1.00
0.1	0.951	0.905	0.820	0.744	0.676	0.614	0.559	0.509	0.464	0.424	0.387
0.2	0.951	0.906	0.822	0.747	0.681	0.622	0.569	0.521	0.479	0.440	0.406
0.3	0.952	0.906	0.824	0.751	0.686	0.629	0.578	0.533	0.493	0.457	0.425
0.4	0.952	0.907	0.825	0.754	0.691	0.636	0.587	0.544	0.506	0.472	0.442
0.5	0.952	0.907	0.827	0.757	0.696	0.643	0.596	0.555	0.519	0.487	0.459
0.6	0.952	0.907	0.828	0.760	0.701	0.650	0.605	0.566	0.532	0.502	0.476
0.7	0.952	0.908	0.830	0.763	0.706	0.656	0.614	0.577	0.544	0.516	0.492
0.8	0.952	0.908	0.831	0.766	0.710	0.663	0.622	0.587	0.556	0.530	0.507
0.9	0.952	0.909	0.833	0.769	0.715	0.669	0.630	0.597	0.568	0.543	0.522
1.0	0.952	0.909	0.834	0.772	0.719	0.675	0.638	0.606	0.579	0.556	0.536
1.1	0.953	0.910	0.836	0.775	0.724	0.681	0.646	0.616	0.590	0.568	0.550
1.2	0.953	0.910	0.837	0.777	0.728	0.687	0.653	0.625	0.601	0.580	0.563
1.3	0.953	0.910	0.839	0.780	0.732	0.693	0.660	0.633	0.611	0.592	0.576
1.4	0.953	0.911	0.840	0.783	0.736	0.699	0.668	0.642	0.620	0.603	0.588
1.5	0.953	0.911	0.841	0.785	0.741	0.704	0.674	0.650	0.630	0.613	0.600
1.6	0.953	0.912	0.843	0.788	0.745	0.709	0.681	0.658	0.639	0.624	0.611
1.7	0.953	0.912	0.844	0.791	0.748	0.715	0.688	0.666	0.648	0.634	0.622
1.8	0.953	0.913	0.845	0.793	0.752	0.720	0.694	0.673	0.657	0.643	0.632
1.9	0.953	0.913	0.847	0.796	0.756	0.725	0.700	0.681	0.665	0.652	0.642
2.0	0.954	0.913	0.848	0.798	0.760	0.730	0.706	0.688	0.673	0.661	0.652
2.1	0.954	0.914	0.849	0.801	0.763	0.734	0.712	0.694	0.681	0.670	0.661
2.2	0.954	0.914	0.851	0.803	0.767	0.739	0.718	0.701	0.688	0.678	0.670
2.3	0.954	0.915	0.852	0.805	0.770	0.744	0.723	0.707	0.695	0.686	0.678
2.4	0.954	0.915	0.853	0.808	0.774	0.748	0.728	0.714	0.702	0.693	0.687
2.5	0.954	0.915	0.854	0.810	0.777	0.752	0.734	0.720	0.709	0.701	0.694
2.6	0.954	0.916	0.856	0.812	0.780	0.756	0.739	0.726	0.716	0.708	0.702
2.7	0.954	0.916	0.857	0.814	0.783	0.761	0.744	0.731	0.722	0.715	0.709
2.8	0.954	0.916	0.858	0.816	0.786	0.765	0.749	0.737	0.728	0.721	0.716
2.9	0.955	0.917	0.859	0.819	0.789	0.768	0.753	0.742	0.734	0.728	0.723
3.0	0.955	0.917	0.860	0.821	0.792	0.772	0.758	0.747	0.739	0.734	0.729

Table B.11 – (continued)

β \ γ	1.2	1.4	1.6	1.8	2.0	2.3	2.6	2.9	3.2	3.6	4.0
0.1	0.325	0.274	0.233	0.199	0.171	0.139	0.115	0.097	0.084	0.071	0.063
0.2	0.348	0.301	0.263	0.231	0.206	0.177	0.155	0.139	0.127	0.115	0.107
0.3	0.370	0.326	0.291	0.263	0.240	0.213	0.194	0.179	0.168	0.158	0.150
0.4	0.392	0.351	0.319	0.293	0.272	0.249	0.231	0.218	0.208	0.199	0.192
0.5	0.412	0.375	0.346	0.322	0.304	0.282	0.267	0.255	0.246	0.238	0.231
0.6	0.432	0.398	0.371	0.350	0.333	0.314	0.301	0.290	0.283	0.275	0.269
0.7	0.451	0.420	0.396	0.377	0.362	0.345	0.333	0.324	0.317	0.310	0.305
0.8	0.470	0.441	0.419	0.402	0.389	0.374	0.363	0.355	0.349	0.343	0.338
0.9	0.487	0.461	0.441	0.426	0.414	0.401	0.392	0.385	0.380	0.374	0.370
1.0	0.504	0.480	0.463	0.449	0.438	0.427	0.419	0.413	0.408	0.403	0.399
1.1	0.520	0.499	0.483	0.470	0.461	0.451	0.444	0.439	0.434	0.430	0.426
1.2	0.536	0.516	0.502	0.491	0.483	0.474	0.468	0.463	0.459	0.455	0.451
1.3	0.551	0.533	0.520	0.510	0.503	0.495	0.490	0.486	0.482	0.478	0.474
1.4	0.565	0.549	0.537	0.529	0.522	0.515	0.510	0.507	0.503	0.500	0.496
1.5	0.579	0.564	0.553	0.546	0.540	0.534	0.530	0.526	0.523	0.520	0.516
1.6	0.592	0.578	0.569	0.562	0.557	0.552	0.548	0.545	0.542	0.538	0.535
1.7	0.604	0.592	0.584	0.578	0.573	0.568	0.565	0.562	0.559	0.556	0.553
1.8	0.616	0.605	0.598	0.592	0.588	0.584	0.581	0.578	0.575	0.572	0.569
1.9	0.627	0.617	0.611	0.606	0.602	0.599	0.595	0.593	0.590	0.587	0.584
2.0	0.638	0.629	0.623	0.619	0.616	0.612	0.609	0.607	0.605	0.602	0.599
2.1	0.649	0.641	0.635	0.631	0.628	0.625	0.623	0.620	0.618	0.615	0.612
2.2	0.659	0.651	0.646	0.643	0.640	0.637	0.635	0.633	0.630	0.627	0.625
2.3	0.668	0.661	0.657	0.654	0.651	0.649	0.646	0.644	0.642	0.639	0.636
2.4	0.677	0.671	0.667	0.664	0.662	0.660	0.657	0.655	0.653	0.650	0.648
2.5	0.686	0.680	0.677	0.674	0.672	0.670	0.668	0.665	0.663	0.661	0.658
2.6	0.694	0.689	0.686	0.683	0.682	0.679	0.677	0.675	0.673	0.671	0.668
2.7	0.702	0.697	0.694	0.692	0.691	0.688	0.686	0.684	0.683	0.680	0.677
2.8	0.710	0.705	0.703	0.701	0.699	0.697	0.695	0.693	0.691	0.689	0.686
2.9	0.717	0.713	0.711	0.709	0.707	0.705	0.703	0.701	0.700	0.697	0.695
3.0	0.724	0.720	0.718	0.716	0.715	0.713	0.711	0.709	0.708	0.705	0.703

B.3.2 *Heat Exchangers with Three Passages of Internal Fluid (Fig. 2.6)*

Table B.12 – Heat exchangers – E and F type – 2 passages of external fluid – Fig. A.10 – Factor ψ

	γ										
β	0.05	0.10	0.20	0.30	0.40	0.50	0.60	0.70	0.80	0.90	1.00
0.1	0.951	0.905	0.820	0.744	0.676	0.614	0.559	0.509	0.465	0.424	0.388
0.2	0.951	0.906	0.822	0.747	0.681	0.622	0.569	0.522	0.479	0.442	0.408
0.3	0.952	0.906	0.824	0.751	0.686	0.629	0.579	0.534	0.494	0.458	0.427
0.4	0.952	0.907	0.825	0.754	0.691	0.636	0.588	0.545	0.508	0.474	0.445
0.5	0.952	0.907	0.827	0.757	0.696	0.644	0.597	0.557	0.521	0.490	0.462
0.6	0.952	0.908	0.828	0.760	0.701	0.650	0.606	0.568	0.534	0.505	0.479
0.7	0.952	0.908	0.830	0.763	0.706	0.657	0.615	0.578	0.547	0.519	0.495
0.8	0.952	0.908	0.831	0.766	0.711	0.664	0.623	0.589	0.559	0.533	0.511
0.9	0.952	0.909	0.833	0.769	0.716	0.670	0.632	0.599	0.571	0.547	0.526
1.0	0.952	0.909	0.834	0.772	0.720	0.676	0.640	0.608	0.582	0.560	0.541
1.1	0.953	0.910	0.836	0.775	0.724	0.682	0.647	0.618	0.593	0.572	0.555
1.2	0.953	0.910	0.837	0.778	0.729	0.688	0.655	0.627	0.604	0.584	0.568
1.3	0.953	0.911	0.839	0.780	0.733	0.694	0.662	0.636	0.614	0.596	0.581
1.4	0.953	0.911	0.840	0.783	0.737	0.700	0.669	0.644	0.624	0.607	0.593
1.5	0.953	0.911	0.841	0.786	0.741	0.705	0.676	0.653	0.633	0.618	0.605
1.6	0.953	0.912	0.843	0.789	0.745	0.711	0.683	0.661	0.643	0.628	0.617
1.7	0.953	0.912	0.844	0.791	0.749	0.716	0.690	0.669	0.652	0.638	0.627
1.8	0.953	0.913	0.846	0.794	0.753	0.721	0.696	0.676	0.660	0.648	0.638
1.9	0.953	0.913	0.847	0.796	0.757	0.726	0.702	0.683	0.669	0.657	0.648
2.0	0.954	0.913	0.848	0.799	0.761	0.731	0.708	0.691	0.677	0.666	0.658
2.1	0.954	0.914	0.850	0.801	0.764	0.736	0.714	0.697	0.685	0.675	0.667
2.2	0.954	0.914	0.851	0.803	0.768	0.741	0.720	0.704	0.692	0.683	0.676
2.3	0.954	0.915	0.852	0.806	0.771	0.745	0.725	0.711	0.699	0.691	0.685
2.4	0.954	0.915	0.853	0.808	0.775	0.750	0.731	0.717	0.706	0.699	0.693
2.5	0.954	0.915	0.855	0.810	0.778	0.754	0.736	0.723	0.713	0.706	0.701
2.6	0.954	0.916	0.856	0.813	0.781	0.758	0.741	0.729	0.720	0.713	0.708
2.7	0.954	0.916	0.857	0.815	0.784	0.762	0.746	0.734	0.726	0.720	0.716
2.8	0.954	0.917	0.858	0.817	0.787	0.766	0.751	0.740	0.732	0.726	0.723
2.9	0.955	0.917	0.859	0.819	0.791	0.770	0.756	0.745	0.738	0.733	0.729
3.0	0.955	0.917	0.861	0.821	0.794	0.774	0.760	0.750	0.744	0.739	0.736

Table B.12 – (continued)

β	γ 1.2	1.4	1.6	1.8	2.0	2.3	2.6	2.9	3.2	3.6	4.0
0.1	0.326	0.276	0.235	0.202	0.175	0.143	0.121	0.104	0.092	0.081	0.075
0.2	0.350	0.304	0.266	0.236	0.212	0.185	0.165	0.151	0.141	0.133	0.128
0.3	0.373	0.331	0.297	0.270	0.248	0.224	0.207	0.196	0.188	0.182	0.180
0.4	0.395	0.356	0.326	0.302	0.283	0.262	0.248	0.238	0.233	0.229	0.228
0.5	0.417	0.381	0.354	0.332	0.316	0.298	0.286	0.279	0.275	**0.273**	0.273
0.6	0.437	0.405	0.380	0.361	0.347	0.332	0.322	0.317	**0.314**	0.314	0.316
0.7	0.457	0.427	0.405	0.389	0.376	0.364	0.356	0.353	**0.351**	0.352	0.355
0.8	0.476	0.449	0.429	0.415	0.404	0.394	0.388	**0.386**	0.386	0.388	0.392
0.9	0.494	0.470	0.452	0.440	0.431	0.422	0.418	**0.417**	0.418	0.421	0.425
1.0	0.511	0.489	0.474	0.463	0.456	0.449	**0.446**	0.446	0.448	0.452	0.456
1.1	0.527	0.508	0.495	0.485	0.479	0.474	**0.473**	0.473	0.476	0.480	0.485
1.2	0.543	0.526	0.514	0.506	0.501	0.498	**0.497**	0.499	0.501	0.506	0.511
1.3	0.558	0.543	0.533	0.526	0.522	**0.520**	0.520	0.522	0.525	0.530	0.535
1.4	0.573	0.559	0.550	0.545	0.542	**0.540**	0.541	0.544	0.547	0.552	0.558
1.5	0.586	0.574	0.567	0.562	0.560	**0.559**	0.561	0.564	0.567	0.572	0.578
1.6	0.600	0.589	0.582	0.579	**0.577**	0.577	0.579	0.582	0.586	0.591	0.597
1.7	0.612	0.603	0.597	0.595	**0.594**	0.594	0.597	0.600	0.603	0.609	0.614
1.8	0.624	0.616	0.611	**0.609**	0.609	0.610	0.613	0.616	0.620	0.625	0.630
1.9	0.636	0.628	0.625	**0.623**	0.623	0.625	0.628	0.631	0.635	0.640	0.645
2.0	0.647	0.640	0.637	**0.636**	0.637	0.639	0.641	0.645	0.649	0.653	0.658
2.1	0.657	0.652	0.649	**0.649**	0.649	0.651	0.655	0.658	0.661	0.666	0.671
2.2	0.667	0.662	0.661	**0.660**	0.661	0.664	0.667	0.670	0.674	0.678	0.682
2.3	0.677	0.673	**0.671**	0.671	0.672	0.675	0.678	0.681	0.685	0.689	0.693
2.4	0.686	0.682	**0.681**	0.682	0.683	0.686	0.689	0.692	0.695	0.699	0.703
2.5	0.694	0.692	**0.691**	0.692	0.693	0.696	0.699	0.702	0.705	0.709	0.713
2.6	0.703	0.701	**0.700**	0.701	0.702	0.705	0.708	0.711	0.714	0.718	0.722
2.7	0.711	**0.709**	0.709	0.710	0.711	0.714	0.717	0.720	0.723	0.727	0.730
2.8	0.718	**0.717**	0.717	0.718	0.720	0.723	0.726	0.728	0.731	0.735	0.738
2.9	0.726	**0.725**	0.725	0.726	0.728	0.731	0.733	0.736	0.739	0.742	0.745
3.0	0.733	**0.732**	0.732	0.734	0.735	0.738	0.741	0.744	0.746	0.749	-----

Minimum values of ψ are in bold type

Table B.13 – Heat exchangers – G and H type – 2 passages of external fluid – Fig. A.11 – Factor ψ

	γ										
β	0.05	0.10	0.20	0.30	0.40	0.50	0.60	0.70	0.80	0.90	1.00
0.1	0.951	0.905	0.820	0.744	0.676	0.614	0.559	0.509	0.464	0.423	0.387
0.2	0.951	0.906	0.822	0.747	0.681	0.621	0.568	0.521	0.478	0.440	0.406
0.3	0.952	0.906	0.824	0.751	0.686	0.629	0.578	0.532	0.492	0.456	0.424
0.4	0.952	0.907	0.825	0.754	0.691	0.636	0.587	0.544	0.505	0.471	0.441
0.5	0.952	0.907	0.827	0.757	0.696	0.643	0.596	0.555	0.518	0.486	0.458
0.6	0.952	0.907	0.828	0.760	0.701	0.649	0.605	0.565	0.531	0.501	0.474
0.7	0.952	0.908	0.830	0.763	0.796	0.656	0.613	0.576	0.543	0.515	0.490
0.8	0.952	0.908	0.831	0.766	0.710	0.662	0.621	0.586	0.555	0.528	0.505
0.9	0.952	0.909	0.833	0.769	0.715	0.669	0.629	0.596	0.567	0.542	0.520
1.0	0.952	0.909	0.834	0.772	0.719	0.675	0.637	0.605	0.578	0.554	0.534
1.1	0.953	0.910	0.836	0.774	0.724	0.681	0.645	0.614	0.589	0.566	0.548
1.2	0.953	0.910	0.837	0.777	0.728	0.687	0.652	0.623	0.599	0.578	0.561
1.3	0.953	0.910	0.838	0.780	0.732	0.692	0.660	0.632	0.609	0.590	0.573
1.4	0.953	0.911	0.840	0.783	0.736	0.698	0.667	0.641	0.619	0.601	0.585
1.5	0.953	0.911	0.841	0.785	0.740	0.703	0.673	0.649	0.628	0.611	0.597
1.6	0.953	0.912	0.843	0.788	0.744	0.709	0.680	0.657	0.637	0.622	0.608
1.7	0.953	0.912	0.844	0.790	0.748	0.714	0.687	0.664	0.646	0.631	0.619
1.8	0.953	0.913	0.845	0.793	0.752	0.719	0.693	0.672	0.655	0.641	0.629
1.9	0.953	0.913	0.847	0.795	0.756	0.724	0.699	0.679	0.663	0.650	0.639
2.0	0.954	0.913	0.848	0.798	0.759	0.729	0.705	0.686	0.671	0.659	0.649
2.1	0.954	0.914	0.849	0.800	0.763	0.734	0.711	0.693	0.679	0.667	0.658
2.2	0.954	0.914	0.851	0.803	0.766	0.738	0.717	0.700	0.686	0.676	0.667
2.3	0.954	0.915	0.852	0.805	0.770	0.743	0.722	0.706	0.693	0.683	0.676
2.4	0.954	0.915	0.853	0.807	0.773	0.747	0.727	0.712	0.700	0.691	0.684
2.5	0.954	0.915	0.854	0.810	0.776	0.751	0.733	0.718	0.707	0.698	0.692
2.6	0.954	0.916	0.856	0.812	0.780	0.756	0.738	0.724	0.714	0.705	0.699
2.7	0.954	0.916	0.857	0.814	0.783	0.760	0.743	0.730	0.720	0.712	0.706
2.8	0.954	0.916	0.858	0.816	0.786	0.764	0.747	0.735	0.726	0.719	0.713
2.9	0.955	0.917	0.859	0.818	0.789	0.768	0.752	0.740	0.732	0.725	0.720
3.0	0.955	0.917	0.860	0.820	0.792	0.771	0.757	0.746	0.737	0.731	0.726

Table B.13 – (continued)

β \ γ	1.2	1.4	1.6	1.8	2.0	2.3	2.6	2.9	3.2	3.6	4.0
0.1	0.324	0.273	0.232	0.198	0.170	0.137	0.113	0.094	0.081	0.067	0.058
0.2	0.347	0.299	0.261	0.229	0.203	0.173	0.150	0.133	0.120	0.107	0.098
0.3	0.369	0.325	0.289	0.260	0.236	0.208	0.187	0.172	0.159	0.147	0.137
0.4	0.390	0.349	0.316	0.289	0.268	0.242	0.223	0.209	0.197	0.185	0.176
0.5	0.410	0.372	0.342	0.318	0.298	0.275	0.258	0.244	0.234	0.223	0.214
0.6	0.430	0.395	0.367	0.345	0.327	0.306	0.291	0.278	0.269	0.258	0.250
0.7	0.449	0.417	0.391	0.371	0.355	0.336	0.322	0.311	0.302	0.292	0.284
0.8	0.467	0.437	0.414	0.396	0.381	0.365	0.352	0.342	0.333	0.324	0.316
0.9	0.484	0.457	0.436	0.420	0.407	0.391	0.380	0.371	0.363	0.354	0.347
1.0	0.501	0.476	0.457	0.442	0.430	0.417	0.406	0.398	0.391	0.383	0.375
1.1	0.517	0.494	0.477	0.464	0.453	0.441	0.431	0.423	0.417	0.409	0.402
1.2	0.533	0.512	0.496	0.484	0.474	0.463	0.454	0.447	0.441	0.434	0.427
1.3	0.547	0.528	0.514	0.503	0.494	0.484	0.476	0.470	0.464	0.457	0.450
1.4	0.561	0.544	0.531	0.521	0.513	0.504	0.497	0.491	0.485	0.478	0.472
1.5	0.575	0.559	0.547	0.538	0.531	0.523	0.516	0.510	0.505	0.499	0.492
1.6	0.588	0.573	0.563	0.555	0.548	0.540	0.534	0.529	0.524	0.517	0.512
1.7	0.600	0.587	0.577	0.570	0.564	0.557	0.551	0.546	0.541	0.535	0.529
1.8	0.612	0.600	0.591	0.584	0.579	0.572	0.567	0.562	0.557	0.552	0.546
1.9	0.623	0.612	0.604	0.598	0.593	0.587	0.582	0.577	0.573	0.567	0.562
2.0	0.634	0.624	0.617	0.611	0.606	0.601	0.596	0.591	0.587	0.582	0.577
2.1	0.645	0.635	0.629	0.623	0.619	0.614	0.609	0.605	0.601	0.596	0.591
2.2	0.655	0.646	0.640	0.635	0.631	0.626	0.622	0.617	0.614	0.609	0.604
2.3	0.664	0.656	0.650	0.646	0.642	0.637	0.633	0.629	0.626	0.621	0.617
2.4	0.673	0.666	0.661	0.656	0.653	0.648	0.644	0.641	0.637	0.633	0.629
2.5	0.682	0.675	0.670	0.666	0.663	0.659	0.655	0.651	0.648	0.644	0.640
2.6	0.690	0.684	0.679	0.676	0.673	0.668	0.665	0.661	0.658	0.654	0.651
2.7	0.698	0.692	0.688	0.685	0.682	0.678	0.674	0.671	0.668	0.664	0.661
2.8	0.706	0.700	0.696	0.693	0.690	0.687	0.683	0.680	0.677	0.674	0.670
2.9	0.713	0.708	0.704	0.701	0.699	0.695	0.692	0.689	0.686	0.682	0.680
3.0	0.720	0.715	0.712	0.709	0.706	0.703	0.700	0.697	0.694	0.691	0.688

Table B.14 – Heat exchangers – E type – 3 passages of external fluid – Fig. A.12 – Factor ψ

β	γ 0.05	0.10	0.20	0.30	0.40	0.50	0.60	0.70	0.80	0.90	1.00
0.1	0.951	0.905	0.820	0.744	0.676	0.614	0.559	0.510	0.465	0.425	0.389
0.2	0.951	0.906	0.822	0.748	0.681	0.622	0.569	0.522	0.480	0.443	0.409
0.3	0.952	0.906	0.824	0.751	0.687	0.630	0.579	0.535	0.495	0.460	0.428
0.4	0.952	0.907	0.825	0.754	0.692	0.637	0.589	0.546	0.509	0.476	0.447
0.5	0.952	0.907	0.827	0.757	0.697	0.644	0.598	0.558	0.523	0.492	0.465
0.6	0.952	0.908	0.828	0.760	0.702	0.651	0.607	0.569	0.536	0.507	0.482
0.7	0.952	0.908	0.830	0.763	0.707	0.658	0.616	0.580	0.549	0.522	0.499
0.8	0.952	0.908	0.831	0.766	0.711	0.665	0.625	0.590	0.561	0.536	0.515
0.9	0.952	0.909	0.833	0.769	0.716	0.671	0.633	0.601	0.573	0.550	0.530
1.0	0.952	0.909	0.834	0.772	0.721	0.677	0.641	0.610	0.585	0.563	0.545
1.1	0.953	0.910	0.836	0.775	0.725	0.684	0.649	0.620	0.596	0.576	0.559
1.2	0.953	0.910	0.837	0.778	0.729	0.690	0.657	0.629	0.607	0.588	0.572
1.3	0.953	0.911	0.839	0.781	0.734	0.695	0.664	0.638	0.617	0.600	0.585
1.4	0.953	0.911	0.840	0.784	0.738	0.701	0.671	0.647	0.627	0.611	0.598
1.5	0.953	0.911	0.842	0.786	0.742	0.707	0.678	0.655	0.637	0.622	0.610
1.6	0.953	0.912	0.843	0.789	0.746	0.712	0.685	0.663	0.646	0.632	0.621
1.7	0.953	0.912	0.844	0.792	0.750	0.717	0.692	0.671	0.655	0.642	0.632
1.8	0.953	0.913	0.846	0.794	0.754	0.723	0.698	0.679	0.663	0.652	0.642
1.9	0.953	0.913	0.847	0.797	0.758	0.728	0.704	0.686	0.672	0.661	0.653
2.0	0.954	0.913	0.848	0.799	0.761	0.732	0.710	0.693	0.680	0.670	0.662
2.1	0.954	0.914	0.850	0.802	0.765	0.737	0.716	0.700	0.688	0.678	0.671
2.2	0.954	0.914	0.851	0.804	0.769	0.742	0.722	0.707	0.695	0.687	0.680
2.3	0.954	0.915	0.852	0.806	0.772	0.747	0.727	0.713	0.702	0.695	0.689
2.4	0.954	0.915	0.854	0.809	0.775	0.751	0.733	0.719	0.709	0.702	0.697
2.5	0.954	0.915	0.855	0.811	0.779	0.755	0.738	0.725	0.716	0.710	0.705
2.6	0.954	0.916	0.856	0.813	0.782	0.760	0.743	0.731	0.723	0.717	0.712
2.7	0.954	0.916	0.857	0.815	0.785	0.764	0.748	0.737	0.729	0.723	0.720
2,8	0.954	0.917	0.858	0.818	0.788	0.768	0.753	0.742	0.735	0.730	0.727
2.9	0.955	0.917	0.860	0.820	0.791	0.772	0.758	0.748	0.741	0.736	0.733
3.0	0.955	0.917	0.861	0.822	0.794	0.775	0.762	0.753	0.747	0.742	0.740

Table B.14 – (continued)

β	γ 1.2	1.4	1.6	1.8	2.0	2.3	2.6	2.9	3.2	3.6	4.0
0.1	0.327	0.277	0.237	0.204	0.177	0.146	0.124	0.108	0.097	0.087	0.080
0.2	0.352	0.306	0.270	0.240	0.217	0.190	0.172	0.159	0.150	0.142	0.139
0.3	0.376	0.334	0.301	0.275	0.255	0.232	0.216	0.206	0.199	0.195	0.193
0.4	0.399	0.361	0.331	0.308	0.290	0.271	0.259	0.251	0.246	**0.243**	0.243
0.5	0.420	0.386	0.360	0.340	0.324	0.308	0.298	0.292	0.289	**0.288**	0.290
0.6	0.441	0.410	0.387	0.369	0.356	0.343	0.335	0.331	**0.330**	0.330	0.333
0.7	0.461	0.433	0.413	0.397	0.386	0.376	0.370	**0.367**	0.367	0.369	0.372
0.8	0.481	0.455	0.437	0.424	0.415	0.406	0.402	**0.401**	0.401	0.404	0.408
0.9	0.499	0.476	0.460	0.449	0.441	0.435	**0.432**	0.432	0.433	0.437	0.441
1.0	0.516	0.496	0.482	0.473	0.466	0.462	**0.460**	0.461	0.463	0.466	0.471
1.1	0.533	0.515	0.503	0.495	0.490	**0.486**	0.486	0.487	0.490	0.494	0.498
1.2	0.549	0.533	0.522	0.516	0.512	**0.510**	0.510	0.512	0.515	0.519	0.523
1.3	0.564	0.550	0.541	0.536	0.533	**0.531**	0.532	0.535	0.537	0.542	0.546
1.4	0.578	0.566	0.558	0.554	0.552	**0.551**	0.553	0.555	0.558	0.563	0.566
1.5	0.592	0.581	0.575	0.571	**0.570**	0.570	0.572	0.575	0.578	0.582	0.585
1.6	0.605	0.596	0.590	0.588	**0.587**	0.588	0.590	0.593	0.596	0.600	0.603
1.7	0.618	0.610	0.605	**0.603**	0.603	0.604	0.607	0.609	0.612	0.616	0.619
1.8	0.630	0.623	0.619	**0.618**	0.618	0.619	0.622	0.625	0.628	0.631	0.634
1.9	0.641	0.635	0.632	**0.631**	0.632	0.634	0.636	0.639	0.642	0.645	0.647
2.0	0.652	0.647	0.645	**0.644**	0.645	0.647	0.650	0.652	0.655	0.658	0.660
2.1	0.663	0.658	**0.656**	0.656	0.657	0.659	0.662	0.665	0.667	0.670	0.672
2.2	0.672	0.669	**0.667**	0.668	0.669	0.671	0.674	0.676	0.678	0.681	0.683
2.3	0.682	0.679	**0.678**	0.678	0.680	0.682	0.684	0.687	0.689	0.691	0.693
2.4	0.691	**0.688**	0.688	0.689	0.690	0.692	0.695	0.697	0.699	0.701	0.703
2.5	0.699	**0.697**	0.697	0.698	0.699	0.702	0.704	0.706	0.708	0.710	0.712
2.6	0.708	**0.706**	0.706	0.707	0.709	0.711	0.713	0.715	0.717	0.719	-----
2.7	0.716	**0.714**	0.715	0.716	0.717	0.719	0.722	0.723	0.725	0.727	-----
2.8	0.723	**0.722**	0.723	0.724	0.725	0.727	0.730	0.731	0.733	-----	-----
2.9	**0.730**	0.730	0.730	0.732	0.733	0.735	0.737	0.739	0.740	-----	-----
3.0	**0.737**	0.737	0.738	0.739	0.740	0.742	0.744	0.746	0.747	-----	-----

Minimum values of ψ are in bold type

Table B.15 – Heat exchangers – G type – 3 passages of external fluid – Fig. A.13 – Factor ψ

	γ										
β	0.05	0.10	0.20	0.30	0.40	0.50	0.60	0.70	0.80	0.90	1.00
0.1	0.951	0.905	0.820	0.744	0.676	0.614	0.558	0.509	0.463	0.423	0.386
0.2	0.951	0.906	0.822	0.747	0.681	0.621	0.568	0.520	0.477	0.439	0.404
0.3	0.952	0.906	0.824	0.750	0.686	0.628	0.577	0.531	0.491	0.454	0.422
0.4	0.952	0.907	0.825	0.754	0.691	0.635	0.586	0.543	0.504	0.469	0.439
0.5	0.952	0.907	0.827	0.757	0.696	0.642	0.595	0.553	0.517	0.484	0.455
0.6	0.952	0.907	0.828	0.760	0.700	0.649	0.604	0.564	0.529	0.498	0.471
0.7	0.952	0.908	0.830	0.763	0.705	0.655	0.612	0.574	0.541	0.512	0.486
0.8	0.952	0.908	0.831	0.766	0.710	0.662	0.620	0.584	0.553	0.525	0.501
0.9	0.952	0.909	0.833	0.768	0.714	0.668	0.628	0.594	0.564	0.538	0.516
1.0	0.952	0.909	0.834	0.771	0.719	0.674	0.636	0.603	0.575	0.551	0.530
1.1	0.953	0.910	0.836	0.774	0.723	0.680	0.643	0.612	0.586	0.563	0.543
1.2	0.953	0.910	0.837	0.777	0.727	0.686	0.651	0.621	0.596	0.575	0.556
1.3	0.953	0.910	0.838	0.780	0.731	0.691	0.658	0.630	0.606	0.586	0.569
1.4	0.953	0.911	0.840	0.782	0.735	0.697	0.665	0.638	0.616	0.597	0.581
1.5	0.953	0.911	0.841	0.785	0.739	0.702	0.672	0.646	0.625	0.607	0.592
1.6	0.953	0.912	0.842	0.787	0.743	0.708	0.678	0.654	0.634	0.618	0.604
1.7	0.953	0.912	0.844	0.790	0.747	0.713	0.685	0.662	0.643	0.627	0.614
1.8	0.953	0.913	0.845	0.793	0.751	0.718	0.691	0.669	0.652	0.637	0.625
1.9	0.953	0.913	0.846	0.795	0.755	0.723	0.697	0.677	0.660	0.646	0.635
2.0	0.954	0.913	0.848	0.797	0.758	0.728	0.703	0.684	0.668	0.655	0.644
2.1	0.954	0.914	0.849	0.800	0.762	0.732	0.709	0.690	0.675	0.663	0.654
2.2	0.954	0.914	0.850	0.802	0.765	0.737	0.715	0.697	0.683	0.672	0.663
2.3	0.954	0.914	0.852	0.805	0.769	0.741	0.720	0.703	0.690	0.680	0.671
2.4	0.954	0.915	0.853	0.807	0.772	0.746	0.725	0.710	0.697	0.687	0.679
2.5	0.954	0.915	0.854	0.809	0.775	0.750	0.731	0.716	0.704	0.695	0.687
2.6	0.954	0.916	0.855	0.811	0.779	0.754	0.736	0.721	0.710	0.702	0.695
2.7	0.954	0.916	0.857	0.813	0.782	0.758	0.741	0.727	0.717	0.709	0.702
2.8	0.954	0.916	0.858	0.816	0.785	0.762	0.745	0.733	0.723	0.715	0.709
2.9	0.955	0.917	0.859	0.818	0.788	0.766	0.750	0.738	0.729	0.722	0.716
3.0	0.955	0.917	0.860	0.820	0.791	0.770	0.755	0.743	0.734	0.728	0.722

Table B.15 – (continued)

β	γ 1.2	1.4	1.6	1.8	2.0	2.3	2.6	2.9	3.2	3.6	4.0
0.1	0.323	0.272	0.230	0.195	0.167	0.134	0.109	0.090	0.076	0.062	0.052
0.2	0.345	0.297	0.257	0.225	0.199	0.167	0.143	0.125	0.111	0.097	0.086
0.3	0.366	0.321	0.284	0.254	0.230	0.200	0.178	0.161	0.147	0.133	0.122
0.4	0.386	0.344	0.310	0.282	0.260	0.233	0.212	0.195	0.182	0.169	0.157
0.5	0.406	0.367	0.336	0.310	0.289	0.264	0.245	0.229	0.217	0.204	0.193
0.6	0.425	0.389	0.360	0.336	0.317	0.294	0.276	0.262	0.251	0.238	0.227
0.7	0.444	0.410	0.383	0.362	0.344	0.323	0.307	0.294	0.283	0.271	0.260
0.8	0.462	0.431	0.406	0.386	0.370	0.351	0.336	0.324	0.314	0.302	0.292
0.9	0.479	0.450	0.428	0.410	0.395	0.377	0.364	0.353	0.343	0.332	0.322
1.0	0.495	0.469	0.448	0.432	0.419	0.403	0.390	0.380	0.371	0.360	0.351
1.1	0.511	0.487	0.468	0.453	0.441	0.426	0.415	0.405	0.397	0.387	0.378
1.2	0.527	0.504	0.487	0.473	0.462	0.449	0.438	0.429	0.422	0.412	0.403
1.3	0.541	0.521	0.505	0.493	0.482	0.470	0.461	0.452	0.445	0.436	0.427
1.4	0.555	0.537	0.522	0.511	0.502	0.490	0.481	0.474	0.467	0.458	0.450
1.5	0.569	0.552	0.538	0.528	0.520	0.509	0.501	0.494	0.487	0.479	0.471
1.6	0.582	0.566	0.554	0.544	0.537	0.527	0.519	0.513	0.506	0.499	0.491
1.7	0.594	0.580	0.569	0.560	0.553	0.544	0.537	0.530	0.524	0.517	0.510
1.8	0.606	0.593	0.583	0.575	0.568	0.560	0.553	0.547	0.542	0.535	0.528
1.9	0.618	0.605	0.596	0.589	0.583	0.575	0.569	0.563	0.558	0.551	0.545
2.0	0.628	0.617	0.608	0.602	0.596	0.589	0.583	0.578	0.573	0.567	0.561
2.1	0.639	0.628	0.620	0.614	0.609	0.603	0.597	0.592	0.587	0.581	0.576
2.2	0.649	0.639	0.632	0.626	0.621	0.615	0.610	0.605	0.601	0.595	0.590
2.3	0.658	0.649	0.643	0.637	0.633	0.627	0.622	0.618	0.613	0.608	0.604
2.4	0.668	0.659	0.653	0.648	0.644	0.638	0.634	0.629	0.626	0.621	0.617
2.5	0.676	0.669	0.663	0.658	0.654	0.649	0.645	0.641	0.637	0.633	0.629
2.6	0.685	0.678	0.672	0.668	0.664	0.659	0.655	0.651	0.648	0.644	0.640
2.7	0.693	0.686	0.681	0.677	0.674	0.669	0.665	0.661	0.658	0.654	0.651
2.8	0.700	0.684	0.690	0.686	0.683	0.678	0.674	0.671	0.668	0.665	0.661
2.9	0.708	0.702	0.698	0.694	0.691	0.687	0.683	0.680	0.677	0.674	0.671
3.0	0.715	0.710	0.705	0.702	0.699	0.695	0.692	0.689	0.686	0.683	0.681

Table B.16 – Heat exchangers – E and F type – 4 passages of external fluid – Fig. A.14 – Factor ψ

	γ										
β	0.05	0.10	0.20	0.30	0.40	0.50	0.60	0.70	0.80	0.90	1.00
0.1	0.951	0.905	0.820	0.744	0.676	0.614	0.559	0.509	0.465	0.424	0.388
0.2	0.951	0.906	0.822	0.747	0.681	0.622	0.569	0.522	0.480	0.442	0.408
0.3	0.952	0.906	0.824	0.751	0.686	0.629	0.579	0.534	0.494	0.458	0.427
0.4	0.952	0.907	0.825	0.754	0.691	0.637	0.588	0.546	0.508	0.475	0.445
0.5	0.952	0,907	0.827	0.757	0.697	0.644	0.597	0.557	0.521	0.490	0.463
0.6	0.952	0.908	0.828	0.760	0.701	0.651	0.606	0.568	0.534	0.505	0.480
0.7	0.952	0.908	0.830	0.763	0.706	0.657	0.615	0.579	0.547	0.520	0.496
0.8	0.952	0.908	0.831	0.766	0.711	0.664	0.624	0.589	0.559	0.534	0.512
0.9	0.952	0.909	0.833	0.769	0.716	0.670	0.632	0.599	0.571	0.547	0.527
1.0	0.952	0.909	0.834	0.772	0.720	0.677	0.640	0.609	0.583	0.560	0.541
1.1	0.953	0.910	0.836	0.775	0.725	0.683	0.648	0.618	0.594	0.573	0.555
1.2	0.953	0.910	0.837	0.778	0.729	0.689	0.655	0.627	0.604	0.585	0.569
1.3	0.953	0.911	0.839	0.781	0.733	0.694	0.663	0.636	0.615	0.597	0.582
1.4	0.953	0.911	0.840	0.783	0.737	0.700	0.670	0.645	0.625	0.608	0.594
1.5	0.953	0.911	0.842	0.786	0.741	0.706	0.677	0.653	0.634	0.619	0.606
1.6	0.953	0.912	0.843	0.789	0.745	0.711	0.683	0.661	0.643	0.629	0.618
1.7	0.953	0.912	0.844	0.791	0.749	0.716	0.690	0.669	0.652	0.639	0.629
1.8	0.953	0.913	0.846	0.794	0.753	0.721	0.696	0.677	0.661	0.649	0.639
1.9	0.953	0.913	0.847	0.796	0.757	0.727	0.703	0.684	0.669	0.658	0.649
2.0	0.954	0.913	0.848	0.799	0.761	0.731	0.709	0.691	0.678	0.667	0.659
2.1	0.954	0.914	0.850	0.801	0.764	0.736	0.715	0.698	0.685	0.676	0.668
2.2	0.954	0.914	0.851	0.804	0.768	0.741	0.720	0.705	0.693	0.684	0.677
2.3	0.954	0.915	0.852	0.806	0.771	0.745	0.726	0.711	0.700	0.692	0.686
2.4	0.954	0.915	0.853	0.808	0.775	0.750	0.731	0.717	0.707	0.700	0.694
2.5	0.954	0.915	0.855	0.810	0.778	0.754	0.737	0.723	0.714	0.707	0.702
2.6	0.954	0.916	0.856	0.813	0.781	0.758	0.742	0.729	0.720	0.714	0.710
2.7	0.954	0.916	0.857	0.815	0.785	0.763	0.747	0.735	0.727	0.721	0.717
2.8	0.954	0.917	0.858	0.817	0.788	0.767	0.751	0.741	0.733	0.728	0.724
2.9	0.955	0.917	0.859	0.819	0.791	0.770	0.756	0.746	0.739	0.734	0.731
3.0	0.955	0.917	0.861	0.821	0.794	0.774	0.761	0.751	0.745	0.740	0.737

Table B.16 – (continued)

β	γ 1.2	1.4	1.6	1.8	2.0	2.3	2.6	2.9	3.2	3.6	4.0
0.1	0.326	0.276	0.235	0.202	0.174	0.143	0.120	0.103	0.091	0.079	0.072
0.2	0.350	0.304	0.267	0.236	0.212	0.184	0.164	0.149	0.139	0.130	0.124
0.3	0.373	0.331	0.297	0.270	0.248	0.224	0.206	0.194	0.186	0.178	0.174
0.4	0.396	0.357	0.326	0.302	0.283	0.262	0.247	0.237	0.230	0.225	0.222
0.5	0.417	0.382	0.354	0.333	0.316	0.298	0.286	0.278	0.273	0.269	0.268
0.6	0.438	0.406	0.381	0.362	0.348	0.332	0.322	0.316	0.313	**0.311**	0.311
0.7	0.457	0.428	0.406	0.390	0.377	0.365	0.357	0.353	**0.350**	0.350	0.351
0.8	0.476	0.450	0.431	0.416	0.406	0.395	0.389	**0.386**	0.386	0.386	0.388
0.9	0.494	0.471	0.454	0.441	0.432	0.424	0.420	**0.418**	0.418	0.420	0.423
1.0	0.512	0.491	0.475	0.465	0.458	0.451	**0.448**	0.448	0.449	0.451	0.454
1.1	0.528	0.509	0.496	0.487	0.481	0.476	**0.475**	0.475	0.477	0.480	0.483
1.2	0.544	0.527	0.516	0.508	0.504	0.500	**0.499**	0.500	0.502	0.506	0.509
1.3	0.559	0.544	0.535	0.528	0.525	**0.522**	0.522	0.524	0.526	0.530	0.534
1.4	0.574	0.561	0.552	0.547	0.544	**0.543**	0.544	0.546	0.548	0.552	0.556
1.5	0.588	0.576	0.569	0.565	0.563	**0.562**	0.564	0.566	0.569	0.572	0.576
1.6	0.601	0.591	0.585	0.581	**0.580**	0.580	0.582	0.585	0.587	0.591	0.594
1.7	0.614	0.605	0.600	0.597	**0.596**	0.597	0.599	0.602	0.605	0.608	0.611
1.8	0.626	0.618	0.614	**0.612**	0.612	0.613	0.615	0.618	0.621	0.624	0.627
1.9	0.637	0.630	0.627	**0.626**	0.626	0.628	0.630	0.633	0.635	0.639	0.642
2.0	0.648	0.642	0.640	**0.639**	0.639	0.641	0.644	0.647	0.649	0.652	0.655
2.1	0.659	0.654	0.652	**0.651**	0.652	0.654	0.657	0.659	0.662	0.665	0.667
2.2	0.669	0.665	**0.663**	0.663	0.664	0.666	0.669	0.671	0.674	0.676	0.679
2.3	0.678	0.675	**0.674**	0.674	0.675	0.677	0.680	0.683	0.685	0.687	0.689
2.4	0.687	0.685	**0.684**	0.684	0.686	0.688	0.691	0.693	0.695	0.697	0.699
2.5	0.696	0.694	**0.693**	0.694	0.696	0.698	0.700	0.703	0.705	0.707	0.709
2.6	0.704	**0.703**	0.703	0.703	0.705	0.707	0.710	0.712	0.714	0.716	0.720
2.7	0.712	**0.711**	0.711	0.712	0.714	0.716	0.718	0.720	0.722	0.724	-----
2.8	0.720	**0.719**	0.719	0.721	0.722	0.724	0.727	0.729	0.730	0.732	-----
2.9	**0.727**	0.727	0.727	0.728	0.730	0.732	0.734	0.736	0.738	-----	-----
3.0	**0.734**	0.734	0.735	0.736	0.737	0.740	0.742	0.743	0.745	-----	-----

Minimum values of ψ are in bold type

Table B.17 – Heat exchangers – G and H type – 4 passages of external fluid – Fig. A.15 – Factor ψ

	γ										
β	0.05	0.10	0.20	0.30	0.40	0.50	0.60	0.70	0.80	0.90	1.00
0.1	0.951	0.905	0.820	0.744	0.676	0.614	0.559	0.509	0.464	0.423	0.387
0.2	0.951	0.906	0.822	0.747	0.681	0.621	0.568	0.521	0.478	0.440	0.405
0.3	0.952	0.906	0.824	0.751	0.686	0.629	0.578	0.532	0.492	0.456	0.423
0.4	0.952	0.907	0.825	0.754	0.691	0.636	0.587	0.544	0.505	0.471	0.441
0.5	0.952	0.907	0.827	0.757	0.696	0.643	0.596	0.555	0.518	0.486	0.458
0.6	0.952	0.907	0.828	0.760	0.701	0.649	0.604	0.565	0.531	0.501	0.474
0.7	0.952	0.908	0.830	0.763	0.705	0.656	0.613	0.576	0.543	0.515	0.490
0.8	0.952	0.908	0.831	0.766	0.710	0.662	0.621	0.586	0.555	0.528	0.505
0.9	0.952	0.909	0.833	0.769	0.715	0.669	0.629	0.595	0.566	0.541	0.520
1.0	0.952	0.909	0.834	0.772	0.719	0.675	0.637	0.605	0.577	0.554	0.534
1.1	0.953	0.910	0.836	0.774	0.723	0.681	0.645	0.614	0.588	0.566	0.547
1.2	0.953	0.910	0.837	0.777	0.728	0.687	0.652	0.623	0.599	0.578	0.560
1.3	0.953	0.910	0.838	0.780	0.732	0.692	0.659	0.632	0.609	0.589	0.573
1.4	0.953	0.911	0.840	0.783	0.736	0.698	0.666	0.640	0.619	0.600	0.585
1.5	0.953	0.911	0.841	0.785	0.740	0.703	0.673	0.648	0.628	0.611	0.597
1.6	0.953	0.912	0.843	0.788	0.744	0.709	0.680	0.656	0.637	0.621	0.608
1.7	0.953	0.912	0.844	0.790	0.748	0.714	0.686	0.664	0.646	0.631	0.619
1.8	0.953	0.913	0.845	0.793	0.752	0.719	0.693	0.672	0.655	0.641	0.629
1.9	0.953	0.913	0.847	0.795	0.755	0.724	0.699	0.679	0.663	0.650	0.639
2.0	0.954	0.913	0.848	0.798	0.759	0.729	0.705	0.686	0.671	0.659	0.649
2.1	0.954	0.914	0.849	0.800	0.763	0.733	0.711	0.693	0.678	0.667	0.658
2.2	0.954	0.914	0.851	0.803	0.766	0.738	0.716	0.699	0.686	0.675	0.667
2.3	0.954	0.915	0.852	0.805	0.770	0.743	0.722	0.706	0.693	0.683	0.675
2.4	0.954	0.915	0.853	0.807	0.773	0.747	0.727	0.712	0.700	0.691	0.683
2.5	0.954	0.915	0.854	0.809	0.776	0.751	0.732	0.718	0.707	0.698	0.691
2.6	0.954	0.916	0.855	0.812	0.779	0.755	0.737	0.724	0.713	0.705	0.699
2.7	0.954	0.916	0.857	0.814	0.783	0.760	0.742	0.729	0.720	0.712	0.706
2.8	0.954	0.916	0.858	0.816	0.786	0.764	0.747	0.735	0.726	0.718	0.713
2.9	0.955	0.917	0.859	0.818	0.789	0.767	0.752	0.740	0.731	0.725	0.720
3.0	0.955	0.917	0.860	0.820	0.792	0.771	0.756	0.745	0.737	0.731	0.726

Table B.17 – (continued)

β	γ 1.2	1.4	1.6	1.8	2.0	2.3	2.6	2.9	3.2	3.6	4.0
0.1	0.324	0.273	0.232	0.198	0.170	0.137	0.114	0.096	0.082	0.070	0.061
0.2	0.347	0.299	0.261	0.229	0.204	0.174	0.152	0.136	0.124	0.113	0.105
0.3	0.369	0.324	0.289	0.260	0.237	0.210	0.190	0.176	0.165	0.155	0.148
0.4	0.390	0.349	0.316	0.290	0.269	0.244	0.227	0.214	0.204	0.195	0.189
0.5	0.410	0.372	0.342	0.319	0.299	0.278	0.262	0.250	0.242	0.234	0.228
0.6	0.430	0.395	0.368	0.346	0.329	0.309	0.295	0.285	0.278	0.271	0.265
0.7	0.449	0.417	0.392	0.372	0.357	0.339	0.327	0.318	0.311	0.305	0.300
0.8	0.467	0.437	0.415	0.397	0.383	0.368	0.357	0.349	0.343	0.337	0.332
0.9	0.484	0.457	0.437	0.421	0.409	0.395	0.385	0.378	0.372	0.367	0.362
1.0	0.501	0.476	0.458	0.444	0.433	0.420	0.412	0.405	0.400	0.394	0.390
1.1	0.517	0.495	0.478	0.465	0.455	0.444	0.436	0.430	0.425	0.420	0.415
1.2	0.532	0.512	0.497	0.485	0.476	0.467	0.459	0.454	0.449	0.444	0.439
1.3	0.547	0.529	0.515	0.504	0.496	0.488	0.481	0.476	0.471	0.466	0.461
1.4	0.561	0.544	0.532	0.522	0.515	0.507	0.501	0.496	0.492	0.486	0.481
1.5	0.575	0.559	0.548	0.540	0.533	0.526	0.520	0.515	0.511	0.505	0.500
1.6	0.588	0.574	0.563	0.556	0.550	0.543	0.538	0.533	0.529	0.523	0.518
1.7	0.600	0.587	0.578	0.571	0.566	0.559	0.554	0.550	0.545	0.540	0.535
1.8	0.612	0.600	0.592	0.585	0.580	0.575	0.570	0.565	0.561	0.556	0.550
1.9	0.623	0.613	0.605	0.599	0.594	0.589	0.584	0.580	0.576	0.570	0.565
2.0	0.634	0.624	0.617	0.612	0.608	0.602	0.598	0.594	0.590	0.584	0.579
2.1	0.645	0.635	0.629	0.624	0.620	0.615	0.611	0.607	0.603	0.597	0.592
2.2	0.654	0.646	0.640	0.635	0.632	0.627	0.623	0.619	0.615	0.610	0.605
2.3	0.664	0.656	0.651	0.646	0.643	0.638	0.634	0.630	0.626	0.622	0.617
2.4	0.673	0.666	0.661	0.657	0.653	0.649	0.645	0.641	0.638	0.633	0.628
2.5	0.681	0.675	0.670	0.666	0.663	0.659	0.655	0.652	0.648	0.643	0.639
2.6	0.690	0.684	0.679	0.676	0.673	0.669	0.665	0.661	0.658	0.654	-----
2.7	0.698	0.692	0.688	0.685	0.682	0.678	0.674	0.671	0.667	0.663	-----
2.8	0.705	0.700	0.696	0.693	0.690	0.686	0.683	0.680	0.676	-----	-----
2.9	0.712	0.708	0.704	0.701	0.698	0.695	0.691	0.688	0.685	-----	-----
3.0	0.719	0.715	0.711	0.709	0.706	0.702	0.699	0.696	0.693	-----	-----

Table B.18 – Heat exchangers – E type – 5 passages of external fluid – Fig. A.16 – Factor ψ

β \ γ	0.05	0.10	0.20	0.30	0.40	0.50	0.60	0.70	0.80	0.90	1.00
0.1	0.951	0.905	0.820	0.744	0.676	0.614	0.559	0.509	0.465	0.425	0.388
0.2	0.951	0.906	0.822	0.748	0.681	0.622	0.569	0.522	0.480	0.442	0.408
0.3	0.952	0.906	0.824	0.751	0.686	0.629	0.579	0.534	0.494	0.459	0.427
0.4	0.952	0.907	0.825	0.754	0.692	0.637	0.588	0.546	0.508	0.475	0.446
0.5	0.952	0.907	0.827	0.757	0.697	0.644	0.598	0.557	0.522	0.491	0.464
0.6	0.952	0.908	0.828	0.760	0.702	0.651	0.607	0.568	0.535	0.506	0.481
0.7	0.952	0.908	0.830	0.763	0.706	0.658	0.615	0.579	0.548	0.521	0.497
0.8	0.952	0.908	0.831	0.766	0.711	0.664	0.624	0.590	0.560	0.535	0.513
0.9	0.952	0.909	0.833	0.769	0.716	0.671	0.632	0.600	0.572	0.548	0.528
1.0	0.952	0.909	0.834	0.772	0.720	0.677	0.640	0.609	0.583	0.561	0.543
1.1	0.953	0.910	0.836	0.775	0.725	0.683	0.648	0.619	0.595	0.574	0.557
1.2	0.953	0.910	0.837	0.778	0.729	0.689	0.656	0.628	0.605	0.586	0.570
1.3	0.953	0.911	0.839	0.781	0.733	0.695	0.663	0.637	0.616	0.598	0.583
1.4	0.953	0.911	0.840	0.783	0.738	0.701	0.670	0.646	0.626	0.609	0.596
1.5	0.953	0.911	0.842	0.786	0.742	0.706	0.677	0.654	0.635	0.620	0.608
1.6	0.953	0.912	0.843	0.789	0.746	0.711	0.684	0.662	0.644	0.630	0.619
1.7	0.953	0.912	0.844	0.791	0.750	0.717	0.691	0.670	0.653	0.640	0.630
1.8	0.953	0.913	0.846	0.794	0.754	0.722	0.697	0.678	0.662	0.650	0.641
1.9	0.953	0.913	0.847	0.796	0.757	0.727	0.703	0.685	0.670	0.659	0.651
2.0	0.954	0.913	0.848	0.799	0.761	0.732	0.709	0.692	0.679	0.668	0.660
2.1	0.954	0.914	0.850	0.801	0.765	0.737	0.715	0.699	0.686	0.677	0.670
2.2	0.954	0.914	0.851	0.804	0.768	0.741	0.721	0.706	0.694	0.685	0.679
2.3	0.954	0.915	0.852	0.806	0.772	0.746	0.727	0.712	0.701	0.693	0.687
2.4	0.954	0.915	0.853	0.808	0.775	0.750	0.732	0.718	0.708	0.701	0.695
2.5	0.954	0.915	0.855	0.811	0.778	0.755	0.737	0.724	0.715	0.708	0.703
2.6	0.954	0.916	0.856	0.813	0.782	0.759	0.742	0.730	0.721	0.715	0.711
2.7	0.954	0.916	0.857	0.815	0.785	0.763	0.747	0.736	0.728	0.722	0.718
2.8	0.954	0.917	0.858	0.817	0.788	0.767	0.752	0.741	0.734	0.729	0.725
2.9	0.955	0.917	0.860	0.819	0.791	0.771	0.757	0.747	0.740	0.735	0.732
3.0	0.955	0.917	0.861	0.821	0.794	0.775	0.761	0.752	0.745	0.741	0.738

Table B.18 – (continued)

	γ										
β	1.2	1.4	1.6	1.8	2.0	2.3	2.6	2.9	3.2	3.6	4.0
0.1	0.327	0.276	0.235	0.202	0.175	0.144	0.122	0.105	0.093	0.082	0.075
0.2	0.351	0.305	0.268	0.238	0.214	0.187	0.167	0.153	0.143	0.134	0.129
0.3	0.374	0.332	0.299	0.272	0.251	0.227	0.210	0.199	0.191	0.185	0.182
0.4	0.397	0.358	0.328	0.305	0.286	0.266	0.252	0.242	0.237	0.232	0.231
0.5	0.418	0.384	0.356	0.336	0.319	0.302	0.291	0.284	0.280	**0.277**	0.277
0.6	0.439	0.408	0.383	0.365	0.351	0.337	0.328	0.322	0.320	**0.319**	0.320
0.7	0.459	0.430	0.409	0.393	0.381	0.369	0.362	0.359	**0.358**	0.358	0.360
0.8	0.478	0.452	0.433	0.419	0.410	0.400	0.395	**0.393**	0.393	0.394	0.397
0.9	0.496	0.473	0.456	0.445	0.436	0.429	0.425	**0.424**	0.425	0.428	0.431
1.0	0.514	0.493	0.478	0.468	0.461	0.456	**0.454**	0.454	0.455	0.458	0.462
1.1	0.530	0.512	0.499	0.491	0.485	0.481	**0.480**	0.481	0.483	0.486	0.490
1.2	0.546	0.530	0.519	0.512	0.507	0.505	**0.504**	0.506	0.508	0.512	0.516
1.3	0.561	0.547	0.537	0.532	0.528	**0.527**	0.527	0.529	0.532	0.536	0.540
1.4	0.576	0.563	0.555	0.550	0.548	**0.547**	0.548	0.551	0.553	0.557	0.561
1.5	0.590	0.578	0.572	0.568	**0.566**	0.566	0.568	0.570	0.573	0.577	0.581
1.6	0.603	0.593	0.587	0.584	**0.583**	0.584	0.586	0.589	0.592	0.596	0.599
1.7	0.616	0.607	0.602	**0.600**	0.600	0.601	0.603	0.606	0.609	0.612	0.616
1.8	0.628	0.620	0.616	**0.615**	0.615	0.616	0.619	0.622	0.624	0.628	0.631
1.9	0.639	0.633	0.630	**0.629**	0.629	0.631	0.633	0.636	0.639	0.642	0.645
2.0	0.650	0.645	**0.642**	0.642	0.642	0.644	0.647	0.650	0.652	0.655	0.658
2.1	0.660	0.656	**0.654**	0.654	0.655	0.657	0.660	0.662	0.665	0.668	0.670
2.2	0.670	0.667	**0.665**	0.665	0.666	0.669	0.671	0.674	0.676	0.679	0.681
2.3	0.680	0.677	**0.676**	0.676	0.677	0.680	0.683	0.685	0.687	0.690	0.691
2.4	0.689	0.686	**0.686**	0.687	0.688	0.690	0.693	0.695	0.697	0.699	0.701
2.5	0.698	0.696	**0.695**	0.696	0.698	0.700	0.703	0.705	0.707	0.709	0.710
2.6	0.706	**0.704**	0.704	0.705	0.707	0.709	0.712	0.714	0.716	0.717	-----
2.7	0.714	**0.713**	0.713	0.714	0.716	0.718	0.720	0.722	0.724	0.726	-----
2.8	0.722	**0.721**	0.721	0.722	0.724	0.726	0.728	0.730	0.732	-----	-----
2.9	0.729	**0.728**	0.729	0.730	0.732	0.734	0.736	0.738	0.739	-----	-----
3.0	0.736	**0.735**	0.736	0.738	0.739	0.741	0.743	0.745	-----	-----	-----

Minimum values of ψ are in bold type

Table B.19 – Heat exchangers – G type – 5 passages of external fluid – Fig. A.17 – Factor ψ

β	γ 0.05	0.10	0.20	0.30	0.40	0.50	0.60	0.70	0.80	0.90	1.00
0.1	0.951	0.905	0.820	0.744	0.676	0.614	0.559	0.509	0.464	0.423	0.387
0.2	0.951	0.906	0.822	0.747	0.681	0.621	0.568	0.521	0.478	0.439	0.405
0.3	0.952	0.906	0.824	0.751	0.686	0.628	0.577	0.532	0.491	0.455	0.423
0.4	0.952	0.907	0.825	0.754	0.691	0.635	0.587	0.543	0.505	0.470	0.440
0.5	0.952	0.907	0.827	0.757	0.696	0.642	0.595	0.554	0.518	0.485	0.457
0.6	0.952	0.907	0.828	0.760	0.701	0.649	0.604	0.565	0.530	0.500	0.473
0.7	0.952	0.908	0.830	0.763	0.705	0.656	0.613	0.575	0.542	0.514	0.488
0.8	0.952	0.908	0.831	0.766	0.710	0.662	0.621	0.585	0.554	0.527	0.503
0.9	0.952	0.909	0.833	0.769	0.714	0.668	0.629	0.595	0.565	0.540	0.518
1.0	0.952	0.909	0.834	0.771	0.719	0.674	0.636	0.604	0.577	0.553	0.532
1.1	0.953	0.910	0.836	0.774	0.723	0.680	0.644	0.613	0.587	0.565	0.546
1.2	0.953	0.910	0.837	0.777	0.727	0.686	0.652	0.622	0.598	0.577	0.559
1.3	0.953	0.910	0.838	0.780	0.732	0.692	0.659	0.631	0.608	0.588	0.571
1.4	0.953	0.911	0.840	0.782	0.736	0.697	0.666	0.639	0.617	0.599	0.583
1.5	0.953	0.911	0.841	0.785	0.740	0.703	0.673	0.648	0.627	0.610	0.595
1.6	0.953	0.912	0.843	0.788	0.744	0.708	0.679	0.656	0.636	0.620	0.606
1.7	0.953	0.912	0.844	0.790	0.748	0.713	0.686	0.663	0.645	0.630	0.617
1.8	0.953	0.913	0.845	0.793	0.751	0.718	0.692	0.671	0.653	0.639	0.628
1.9	0.953	0.913	0.847	0.795	0.755	0.723	0.698	0.678	0.662	0.648	0.637
2.0	0.954	0.913	0.848	0.798	0.759	0.728	0.704	0.685	0.670	0.657	0.647
2.1	0.954	0.914	0.849	0.800	0.762	0.733	0.710	0.692	0.677	0.666	0.656
2.2	0.954	0.914	0.850	0.802	0.766	0.738	0.716	0.698	0.685	0.674	0.665
2.3	0.954	0.915	0.852	0.805	0.769	0.742	0.721	0.705	0.692	0.682	0.674
2.4	0.954	0.915	0.853	0.807	0.773	0.746	0.726	0.711	0.699	0.689	0.682
2.5	0.954	0.915	0.854	0.809	0.776	0.751	0.732	0.717	0.706	0.697	0.690
2.6	0.954	0.916	0.855	0.812	0.779	0.755	0.737	0.723	0.712	0.704	0.697
2.7	0.954	0.916	0.857	0.814	0.782	0.759	0.742	0.728	0.718	0.711	0.705
2.8	0.954	0.916	0.858	0.816	0.785	0.763	0.746	0.734	0.724	0.717	0.711
2.9	0.955	0.917	0.859	0.818	0.788	0.767	0.751	0.739	0.730	0.724	0.718
3.0	0.955	0.917	0.860	0.820	0.791	0.771	0.756	0.744	0.736	0.730	0.725

Table B.19 – (continued)

β \ γ	1.2	1.4	1.6	1.8	2.0	2.3	2.6	2.9	3.2	3.6	4.0
0.1	0.324	0.273	0.231	0.197	0.169	0.136	0.112	0.094	0.080	0.067	0.058
0.2	0.346	0.298	0.259	0.228	0.202	0.172	0.149	0.132	0.120	0.107	0.098
0.3	0.368	0.323	0.287	0.258	0.234	0.206	0.186	0.170	0.159	0.147	0.138
0.4	0.388	0.347	0.314	0.287	0.265	0.240	0.221	0.207	0.197	0.186	0.178
0.5	0.409	0.370	0.340	0.315	0.296	0.273	0.256	0.243	0.233	0.223	0.216
0.6	0.428	0.393	0.365	0.342	0.325	0.304	0.289	0.277	0.268	0.259	0.252
0.7	0.447	0.414	0.389	0.368	0.352	0.334	0.320	0.310	0.301	0.293	0.286
0.8	0.465	0.435	0.412	0.393	0.379	0.362	0.350	0.340	0.333	0.325	0.318
0.9	0.482	0.455	0.433	0.417	0.404	0.389	0.378	0.369	0.362	0.354	0.348
1.0	0.499	0.474	0.454	0.439	0.427	0.414	0.404	0.396	0.390	0.382	0.375
1.1	0.515	0.492	0.474	0.461	0.450	0.438	0.429	0.421	0.415	0.408	0.401
1.2	0.530	0.509	0.493	0.481	0.471	0.460	0.452	0.445	0.439	0.432	0.425
1.3	0.545	0.526	0.511	0.500	0.491	0.481	0.474	0.467	0.462	0.455	0.448
1.4	0.559	0.541	0.528	0.518	0.510	0.501	0.494	0.488	0.482	0.476	0.469
1.5	0.573	0.556	0.544	0.535	0.528	0.520	0.513	0.507	0.502	0.495	0.489
1.6	0.586	0.571	0.560	0.551	0.545	0.537	0.531	0.525	0.520	0.514	0.507
1.7	0.598	0.584	0.574	0.567	0.561	0.554	0.548	0.542	0.537	0.531	0.525
1.8	0.610	0.597	0.588	0.581	0.576	0.569	0.563	0.558	0.553	0.547	0.541
1.9	0.621	0.610	0.601	0.595	0.590	0.584	0.578	0.573	0.568	0.562	0.556
2.0	0.632	0.621	0.614	0.608	0.603	0.597	0.592	0.587	0.583	0.577	0.571
2.1	0.642	0.633	0.626	0.620	0.616	0.610	0.605	0.600	0.596	0.590	0.585
2.2	0.652	0.643	0.637	0.632	0.628	0.622	0.617	0.613	0.609	0.603	0.598
2.3	0.662	0.654	0.648	0.643	0.639	0.634	0.629	0.625	0.621	0.616	0.611
2.4	0.671	0.663	0.658	0.653	0.650	0.645	0.640	0.636	0.632	0.627	0.623
2.5	0.679	0.672	0.667	0.663	0.660	0.655	0.651	0.647	0.643	0.638	0.634
2.6	0.688	0.681	0.676	0.673	0.669	0.665	0.661	0.657	0.653	0.649	0.645
2.7	0.696	0.690	0.685	0.682	0.678	0.674	0.670	0.666	0.663	0.659	-----
2.8	0.703	0.698	0.693	0.690	0.687	0.683	0.679	0.676	0.672	0.668	-----
2.9	0.711	0.705	0.701	0.698	0.695	0.691	0.688	0.684	0.681	0.677	-----
3.0	0.718	0.713	0.709	0.706	0.703	0.699	0.696	0.693	0.690	-----	-----

B.3.3 Heat Exchangers with Four Passages of Internal Fluid (Fig. 2.7)

Table B.20 – Heat exchangers – I and N type – 2 passages of external fluid – Fig. A.18 – Factor ψ

	γ										
β	0.05	0.10	0.20	0.30	0.40	0.50	0.60	0.70	0.80	0.90	1.00
0.1	0.951	0.905	0.820	0.744	0.676	0.614	0.559	0.509	0.464	0.424	0.388
0.2	0.951	0.906	0.822	0.747	0.681	0.622	0.569	0.521	0.479	0.441	0.407
0.3	0.952	0.906	0.824	0.751	0.686	0.629	0.578	0.533	0.493	0.458	0.426
0.4	0.952	0.907	0.825	0.754	0.691	0.636	0.588	0.545	0.507	0.474	0.444
0.5	0.952	0.907	0.827	0.757	0.696	0.643	0.597	0.556	0.521	0.489	0.461
0.6	0.952	0.908	0.828	0.760	0.701	0.650	0.606	0.567	0.533	0.504	0.478
0.7	0.952	0.908	0.830	0.763	0.706	0.657	0.614	0.578	0.546	0.518	0.494
0.8	0.952	0.908	0.831	0.766	0.711	0.663	0.623	0.588	0.558	0.532	0.510
0.9	0.952	0.909	0.833	0.769	0.715	0.670	0.631	0.598	0.570	0.546	0.525
1.0	0.952	0.909	0.834	0.772	0.720	0.676	0.639	0.608	0.581	0.559	0.539
1.1	0.953	0.910	0.836	0.775	0.724	0.682	0.647	0.617	0.592	0.571	0.553
1.2	0.953	0.910	0.837	0.778	0.729	0.688	0.654	0.626	0.603	0.583	0.566
1.3	0.953	0.911	0.839	0.780	0.733	0.694	0.662	0.635	0.613	0.595	0.579
1.4	0.953	0.911	0.840	0.783	0.737	0.699	0.669	0.644	0.623	0.606	0.592
1.5	0.953	0.911	0.841	0.786	0.741	0.705	0.676	0.652	0.632	0.616	0.603
1.6	0.953	0.912	0.843	0.788	0.745	0.710	0.682	0.660	0.642	0.627	0.615
1.7	0.953	0.912	0.844	0.791	0.749	0.716	0.689	0.668	0.651	0.637	0.626
1.8	0.953	0.913	0.846	0.794	0.753	0.721	0.695	0.675	0.659	0.646	0.636
1.9	0.953	0.913	0.847	0.796	0.757	0.726	0.702	0.683	0.668	0.656	0.646
2.0	0.954	0.913	0.848	0.798	0.760	0.731	0.708	0.690	0.676	0.665	0.656
2.1	0.954	0.914	0.849	0.801	0.764	0.735	0.713	0.696	0.683	0.673	0.665
2.2	0.954	0.914	0.851	0.803	0.767	0.740	0.719	0.703	0.691	0.681	0.674
2.3	0.954	0.915	0.852	0.806	0.771	0.745	0.725	0.710	0.698	0.689	0.683
2.4	0.954	0.915	0.853	0.808	0.774	0.749	0.730	0.716	0.705	0.697	0.691
2.5	0.954	0.915	0.855	0.810	0.778	0.753	0.735	0.722	0.712	0.704	0.699
2.6	0.954	0.916	0.856	0.812	0.781	0.758	0.740	0.728	0.718	0.712	0.706
2.7	0.954	0.916	0.857	0.815	0.784	0.762	0.745	0.733	0.725	0.718	0.714
2.8	0.954	0.916	0.858	0.817	0.787	0.766	0.750	0.739	0.731	0.725	0.721
2.9	0.955	0.917	0.859	0.819	0.790	0.770	0.755	0.744	0.737	0.731	0.728
3.0	0.955	0.917	0.861	0.821	0.793	0.773	0.759	0.749	0.742	0.737	0.734

Table B.20 – (continued)

β	γ 1.2	1.4	1.6	1.8	2.0	2.3	2.6	2.9	3.2	3.6	4.0
0.1	0.326	0.275	0.234	0.201	0.174	0.142	0.119	0.102	0.090	0.078	0.071
0.2	0.349	0.303	0.265	0.235	0.210	0.182	0.162	0.147	0.137	0.128	0.122
0.3	0.372	0.329	0.295	0.268	0.246	0.221	0.203	0.191	0.182	0.175	0.171
0.4	0.394	0.355	0.324	0.299	0.280	0.258	0.243	0.232	0.226	0.220	0.218
0.5	0.415	0.379	0.351	0.329	0.312	0.293	0.280	0.272	0.267	0.263	0.262
0.6	0.436	0.403	0.377	0.358	0.343	0.326	0.316	0.309	0.305	**0.303**	0.303
0.7	0.455	0.425	0.402	0.385	0.372	0.358	0.349	0.344	0.342	**0.341**	0.342
0.8	0.474	0.447	0.426	0.411	0.400	0.388	0.381	0.377	**0.376**	0.376	0.377
0.9	0.492	0.467	0.449	0.436	0.426	0.416	0.411	0.408	**0.407**	0.408	0.411
1.0	0.509	0.487	0.470	0.459	0.451	0.443	0.439	**0.437**	0.437	0.439	0.442
1.1	0.525	0.505	0.491	0.481	0.474	0.468	0.465	**0.464**	0.464	0.467	0.470
1.2	0.541	0.523	0.510	0.502	0.496	0.491	**0.489**	0.489	0.490	0.493	0.497
1.3	0.556	0.540	0.529	0.521	0.517	0.513	**0.512**	0.512	0.514	0.517	0.521
1.4	0.570	0.556	0.546	0.540	0.536	**0.533**	0.533	0.534	0.536	0.539	0.543
1.5	0.584	0.571	0.563	0.558	0.554	**0.552**	0.553	0.554	0.556	0.560	0.564
1.6	0.597	0.586	0.578	0.574	0.572	**0.570**	0.571	0.573	0.575	0.579	0.583
1.7	0.610	0.600	0.593	0.590	0.588	**0.587**	0.588	0.590	0.593	0.597	0.601
1.8	0.622	0.613	0.607	0.604	**0.603**	0.603	0.605	0.607	0.609	0.614	0.618
1.9	0.633	0.625	0.621	0.618	**0.618**	0.618	0.620	0.622	0.625	0.629	0.633
2.0	0.644	0.637	0.633	0.632	**0.631**	0.632	0.634	0.636	0.639	0.643	0.647
2.1	0.655	0.649	0.645	**0.644**	0.644	0.645	0.647	0.650	0.652	0.656	0.661
2.2	0.665	0.659	0.657	**0.656**	0.656	0.657	0.659	0.662	0.665	0.669	0.673
2.3	0.674	0.670	0.667	**0.667**	0.667	0.669	0.671	0.674	0.677	0.681	0.684
2.4	0.683	0.679	0.678	**0.677**	0.678	0.680	0.682	0.685	0.688	0.691	0.695
2.5	0.692	0.689	**0.687**	0.687	0.688	0.690	0.692	0.695	0.698	0.702	0.705
2.6	0.700	0.697	**0.696**	0.697	0.698	0.700	0.702	0.705	0.707	0.711	0.715
2.7	0.708	0.706	**0.705**	0.706	0.707	0.709	0.711	0.714	0.717	0.720	0.724
2.8	0.716	0.714	**0.714**	0.714	0.715	0.717	0.720	0.722	0.725	0.729	0.732
2.9	0.723	0.722	**0.721**	0.722	0.723	0.725	0.728	0.731	0.733	0.737	0.740
3.0	0.730	**0.729**	0.729	0.730	0.731	0.733	0.736	0.738	0.741	0.744	0.747

Minimum values of ψ are in bold type

Table B.21 – Heat exchangers – L and M type – 2 passages of external fluid – Fig. A.19 – Factor ψ

	γ										
β	0.05	0.10	0.20	0.30	0.40	0.50	0.60	0.70	0.80	0.90	1.00
0.1	0.951	0.905	0.820	0.744	0.676	0.614	0.559	0.509	0.464	0.424	0.387
0.2	0.951	0.906	0.822	0.747	0.681	0.622	0.568	0.521	0.478	0.440	0.406
0.3	0.952	0.906	0.824	0.751	0.686	0.629	0.578	0.533	0.492	0.456	0.424
0.4	0.952	0.907	0.825	0.754	0.691	0.636	0.587	0.544	0.506	0.472	0.442
0.5	0.952	0.907	0.827	0.757	0.696	0.643	0.596	0.555	0.519	0.487	0.459
0.6	0.952	0.908	0.828	0.760	0.701	0.650	0.605	0.566	0.532	0.502	0.475
0.7	0.952	0.908	0.830	0.763	0.706	0.656	0.613	0.576	0.544	0.516	0.491
0.8	0.952	0.908	0.831	0.766	0.710	0.663	0.622	0.586	0.556	0.529	0.506
0.9	0.952	0.909	0.833	0.769	0.715	0.669	0.630	0.596	0.568	0.543	0.521
1.0	0.952	0.909	0.834	0.772	0.719	0.675	0.638	0.606	0.579	0.555	0.535
1.1	0.953	0.910	0.836	0.775	0.724	0.681	0.645	0.615	0.590	0.568	0.549
1.2	0.953	0.910	0.837	0.777	0.728	0.687	0.653	0.624	0.600	0.580	0.562
1.3	0.953	0.910	0.839	0.780	0.732	0.693	0.660	0.633	0.610	0.591	0.575
1.4	0.953	0.911	0.840	0.783	0.736	0.698	0.667	0.641	0.620	0.602	0.587
1.5	0.953	0.911	0.841	0.785	0.740	0.704	0.674	0.650	0.629	0.613	0.599
1.6	0.953	0.912	0.843	0.788	0.744	0.709	0.681	0.658	0.639	0.623	0.610
1.7	0.953	0.912	0.844	0.791	0.748	0.714	0.687	0.665	0.647	0.633	0.621
1.8	0.953	0.913	0.845	0.793	0.752	0.720	0.694	0.673	0.656	0.642	0.631
1.9	0.953	0.913	0.847	0.796	0.756	0.725	0.700	0.680	0.664	0.652	0.641
2.0	0.954	0.913	0.848	0.798	0.759	0.729	0.706	0.687	0.672	0.660	0.651
2.1	0.954	0.914	0.849	0.800	0.763	0.734	0.712	0.694	0.680	0.669	0.660
2.2	0.954	0.914	0.851	0.803	0.767	0.739	0.717	0.701	0.687	0.677	0.669
2.3	0.954	0.915	0.852	0.805	0.770	0.743	0.723	0.707	0.695	0.685	0.678
2.4	0.954	0.915	0.853	0.807	0.773	0.748	0.728	0.713	0.702	0.693	0.686
2.5	0.954	0.915	0.854	0.810	0.777	0.752	0.733	0.719	0.708	0.700	0.694
2.6	0.954	0.916	0.856	0.812	0.780	0.756	0.738	0.725	0.715	0.707	0.701
2.7	0.954	0.916	0.857	0.814	0.783	0.760	0.743	0.731	0.721	0.714	0.708
2.8	0.954	0.916	0.858	0.816	0.786	0.764	0.748	0.736	0.727	0.720	0.715
2.9	0.955	0.917	0.859	0.818	0.789	0.768	0.753	0.742	0.733	0.727	0.722
3.0	0.955	0.917	0.860	0.821	0.792	0.772	0.757	0.747	0.739	0.733	0.728

Table B.21 – (continued)

β	γ 1.2	1.4	1.6	1.8	2.0	2.3	2.6	2.9	3.2	3.6	4.0
0.1	0.325	0.274	0.232	0.198	0.171	0.138	0.114	0.096	0.083	0.070	0.061
0.2	0.348	0.300	0.262	0.231	0.205	0.176	0.154	0.137	0.125	0.112	0.104
0.3	0.370	0.326	0.291	0.262	0.239	0.212	0.192	0.177	0.165	0.154	0.146
0.4	0.391	0.351	0.318	0.292	0.271	0.247	0.228	0.215	0.205	0.194	0.186
0.5	0.412	0.374	0.345	0.321	0.302	0.280	0.264	0.252	0.242	0.233	0.225
0.6	0.431	0.397	0.370	0.349	0.331	0.312	0.297	0.286	0.278	0.269	0.263
0.7	0.450	0.419	0.394	0.375	0.360	0.342	0.329	0.320	0.312	0.304	0.298
0.8	0.469	0.440	0.418	0.400	0.386	0.371	0.359	0.351	0.344	0.337	0.331
0.9	0.486	0.460	0.440	0.424	0.412	0.398	0.388	0.380	0.374	0.367	0.362
1.0	0.503	0.479	0.461	0.447	0.436	0.424	0.415	0.408	0.402	0.396	0.390
1.1	0.519	0.497	0.481	0.468	0.459	0.448	0.440	0.433	0.428	0.423	0.417
1.2	0.535	0.515	0.500	0.489	0.480	0.470	0.463	0.458	0.453	0.447	0.442
1.3	0.550	0.531	0.518	0.508	0.500	0.492	0.485	0.480	0.476	0.471	0.466
1.4	0.564	0.547	0.535	0.526	0.519	0.512	0.506	0.501	0.497	0.492	0.487
1.5	0.578	0.562	0.551	0.543	0.537	0.530	0.525	0.521	0.517	0.512	0.508
1.6	0.591	0.577	0.567	0.560	0.554	0.548	0.543	0.539	0.535	0.531	0.526
1.7	0.603	0.590	0.582	0.575	0.570	0.564	0.560	0.556	0.553	0.548	0.544
1.8	0.615	0.603	0.595	0.590	0.585	0.580	0.576	0.572	0.569	0.565	0.560
1.9	0.626	0.616	0.609	0.603	0.599	0.595	0.591	0.587	0.584	0.580	0.576
2.0	0.637	0.628	0.621	0.616	0.613	0.608	0.605	0.601	0.598	0.594	0.590
2.1	0.647	0.639	0.633	0.629	0.625	0.621	0.618	0.615	0.612	0.608	0.604
2.2	0.657	0.649	0.644	0.640	0.637	0.633	0.630	0.627	0.624	0.620	0.617
2.3	0.667	0.660	0.655	0.651	0.648	0.645	0.642	0.639	0.636	0.632	0.629
2.4	0.676	0.669	0.665	0.662	0.659	0.656	0.653	0.650	0.647	0.644	0.640
2.5	0.684	0.679	0.674	0.671	0.669	0.666	0.663	0.660	0.658	0.654	0.651
2.6	0.693	0.687	0.684	0.681	0.678	0.675	0.673	0.670	0.668	0.664	0.661
2.7	0.701	0.696	0.692	0.690	0.688	0.685	0.682	0.679	0.677	0.674	0.671
2.8	0.708	0.704	0.701	0.698	0.696	0.693	0.691	0.688	0.686	0.683	0.680
2.9	0.716	0.711	0.708	0.706	0.704	0.702	0.699	0.697	0.694	0.691	0.689
3.0	0.722	0.719	0.716	0.714	0.712	0.709	0.707	0.705	0.702	0.700	0.697

Table B.22 – Heat exchangers – M and N type – 3 passages of external fluid – Fig. A.20 –Factor ψ

	γ										
β	0.05	0.10	0.20	0.30	0.40	0.50	0.60	0.70	0.80	0.90	1.00
0.1	0.951	0.905	0.820	0.744	0.676	0.614	0.559	0.509	0.464	0.423	0.387
0.2	0.951	0.906	0.822	0.747	0.681	0.621	0.568	0.521	0.478	0.440	0.405
0.3	0.952	0.906	0.824	0.751	0.686	0.629	0.578	0.532	0.492	0.456	0.423
0.4	0.952	0.907	0.825	0.754	0.691	0.636	0.587	0.544	0.505	0.471	0.441
0.5	0.952	0.907	0.827	0.757	0.696	0.643	0.596	0.554	0.518	0.486	0.457
0.6	0.952	0.907	0.828	0.760	0.701	0.649	0.604	0.565	0.531	0.500	0.474
0.7	0.952	0.908	0.830	0.763	0.705	0.656	0.613	0.576	0.543	0.514	0.490
0.8	0.952	0.908	0.831	0.766	0.710	0.662	0.621	0.586	0.555	0.528	0.505
0.9	0.952	0.909	0.833	0.769	0.715	0.669	0.629	0.595	0.566	0.541	0.519
1.0	0.952	0.909	0.834	0.772	0.719	0.675	0.637	0.605	0.577	0.554	0.534
1.1	0.953	0.910	0.836	0.774	0.723	0.681	0.645	0.614	0.588	0.566	0.547
1.2	0.953	0.910	0.837	0.777	0.728	0.687	0.652	0.623	0.599	0.578	0.561
1.3	0.953	0.910	0.838	0.780	0.732	0.692	0.659	0.632	0.609	0.590	0.573
1.4	0.953	0.911	0.840	0.783	0.736	0.698	0.666	0.640	0.619	0.601	0.585
1.5	0.953	0.911	0.841	0.785	0.740	0.703	0.673	0.649	0.628	0.611	0.597
1.6	0.953	0.912	0.843	0.788	0.744	0.709	0.680	0.657	0.637	0.622	0.609
1.7	0.953	0.912	0.844	0.790	0.748	0.714	0.686	0.664	0.646	0.632	0.619
1.8	0.953	0.913	0.845	0.793	0.752	0.719	0.693	0.672	0.655	0.641	0.630
1.9	0.953	0.913	0.847	0.795	0.755	0.724	0.699	0.679	0.663	0.650	0.640
2.0	0.954	0.913	0.848	0.798	0.759	0.729	0.705	0.686	0.671	0.659	0.650
2.1	0.954	0.914	0.849	0.800	0.763	0.734	0.711	0.693	0.679	0.668	0.659
2.2	0.954	0.914	0.851	0.803	0.766	0.738	0.716	0.700	0.686	0.676	0.668
2.3	0.954	0.915	0.852	0.805	0.770	0.743	0.722	0.706	0.694	0.684	0.677
2.4	0.954	0.915	0.853	0.807	0.773	0.747	0.727	0.712	0.701	0.692	0.685
2.5	0.954	0.915	0.854	0.810	0.776	0.751	0.733	0.718	0.708	0.699	0.693
2.6	0.954	0.916	0.855	0.812	0.780	0.756	0.738	0.724	0.714	0.706	0.700
2.7	0.954	0.916	0.857	0.814	0.783	0.760	0.743	0.730	0.720	0.713	0.708
2.8	0.954	0.916	0.858	0.816	0.786	0.764	0.747	0.735	0.727	0.720	0.715
2.9	0.955	0.917	0.859	0.818	0.789	0.768	0.752	0.741	0.732	0.726	0.722
3.0	0.955	0.917	0.860	0.820	0.792	0.771	0.757	0.746	0.738	0.732	0.728

Table B.22 – (continued)

β	γ 1.2	1.4	1.6	1.8	2.0	2.3	2.6	2.9	3.2	3.6	4.0
0.1	0.324	0.273	0.231	0.197	0.169	0.136	0.112	0.093	0.079	0.066	0.056
0.2	0.347	0.299	0.260	0.229	0.203	0.172	0.149	0.132	0.119	0.105	0.095
0.3	0.368	0.324	0.288	0.259	0.235	0.207	0.186	0.170	0.158	0.145	0.135
0.4	0.389	0.348	0.315	0.289	0.267	0.241	0.222	0.208	0.197	0.185	0.176
0.5	0.410	0.372	0.341	0.317	0.297	0.275	0.258	0.245	0.234	0.224	0.215
0.6	0.429	0.394	0.367	0.345	0.327	0.307	0.291	0.280	0.271	0.261	0.254
0.7	0.448	0.416	0.391	0.371	0.355	0.337	0.324	0.314	0.306	0.298	0.291
0.8	0.467	0.437	0.414	0.396	0.382	0.366	0.355	0.346	0.339	0.332	0.326
0.9	0.484	0.457	0.437	0.421	0.408	0.394	0.384	0.376	0.371	0.364	0.359
1.0	0.501	0.476	0.458	0.444	0.433	0.420	0.412	0.405	0.400	0.395	0.390
1.1	0.517	0.495	0.478	0.465	0.456	0.445	0.438	0.432	0.428	0.423	0.419
1.2	0.533	0.512	0.497	0.486	0.478	0.469	0.462	0.457	0.454	0.449	0.445
1.3	0.548	0.529	0.516	0.506	0.499	0.491	0.485	0.481	0.478	0.474	0.470
1.4	0.562	0.545	0.533	0.525	0.518	0.511	0.506	0.503	0.500	0.496	0.493
1.5	0.576	0.561	0.550	0.542	0.537	0.531	0.527	0.523	0.521	0.518	0.514
1.6	0.589	0.575	0.566	0.559	0.554	0.549	0.545	0.543	0.540	0.537	0.534
1.7	0.601	0.589	0.581	0.575	0.570	0.566	0.563	0.560	0.558	0.555	0.552
1.8	0.613	0.602	0.595	0.590	0.586	0.582	0.579	0.577	0.575	0.572	0.569
1.9	0.625	0.615	0.608	0.604	0.600	0.597	0.595	0.593	0.591	0.588	0.585
2.0	0.636	0.627	0.621	0.617	0.614	0.611	0.609	0.607	0.605	0.603	0.600
2.1	0.646	0.638	0.633	0.630	0.627	0.624	0.622	0.621	0.619	0.616	0.614
2.2	0.656	0.649	0.645	0.641	0.639	0.637	0.635	0.633	0.632	0.629	0.627
2.3	0.666	0.660	0.655	0.653	0.651	0.649	0.647	0.645	0.644	0.641	0.639
2.4	0.675	0.669	0.666	0.663	0.662	0.660	0.658	0.657	0.655	0.653	0.650
2.5	0.684	0.679	0.676	0.673	0.672	0.670	0.669	0.667	0.665	0.663	0.661
2.6	0.693	0.688	0.685	0.683	0.682	0.680	0.678	0.677	0.675	0.673	0.671
2.7	0.701	0.696	0.694	0.692	0.691	0.689	0.688	0.686	0.685	0.683	0.680
2.8	0.708	0.704	0.702	0.701	0.699	0.698	0.697	0.695	0.694	0.691	0.689
2.9	0.716	0.712	0.710	0.709	0.708	0.706	0.705	0.704	0.702	0.700	0.698
3.0	0.723	0.720	0.718	0.717	0.716	0.714	0.713	0.711	0.710	0.708	0.706

Table B.23 – Heat exchamgers – L and M type – 4 passages of external fluid – Fig. A.21 –Factor ψ

	γ										
β	0.05	0.10	0.20	0.30	0.40	0.50	0.60	0.70	0.80	0.90	1.00
0.1	0.951	0.905	0.820	0.744	0.676	0.614	0.559	0.509	0.464	0.424	0.387
0.2	0.951	0.906	0.822	0.747	0.681	0.622	0.569	0.521	0.479	0.441	0.406
0.3	0.952	0.906	0.824	0.751	0.686	0.629	0.578	0.533	0.493	0.457	0.425
0.4	0.952	0.907	0.825	0.754	0.691	0.636	0.587	0.544	0.506	0.473	0.443
0.5	0.952	0.907	0.827	0.757	0.696	0.643	0.596	0.556	0.520	0.488	0.460
0.6	0.952	0.907	0.828	0.760	0.701	0.650	0.605	0.566	0.532	0.503	0.476
0.7	0.952	0.908	0.830	0.763	0.706	0.656	0.614	0.577	0.545	0.517	0.492
0.8	0.952	0.908	0.831	0.766	0.710	0.663	0.622	0.587	0.557	0.531	0.508
0.9	0.952	0.909	0.833	0.769	0.715	0.669	0.630	0.597	0.568	0.544	0.523
1.0	0.952	0.909	0.834	0.772	0.720	0.676	0.638	0.607	0.580	0.557	0.537
1.1	0.953	0.910	0.836	0.775	0.724	0.682	0.646	0.616	0.591	0.569	0.551
1.2	0.953	0.910	0.837	0.777	0.728	0.687	0.653	0.625	0.601	0.581	0.564
1.3	0.953	0.910	0.839	0.780	0.732	0.693	0.661	0.634	0.611	0.593	0.577
1.4	0.953	0.911	0.840	0.783	0.737	0.699	0.668	0.642	0.621	0.604	0.589
1.5	0.953	0.911	0.841	0.786	0.741	0.704	0.675	0.651	0.631	0.614	0.601
1.6	0.953	0.912	0.843	0.788	0.745	0.710	0.681	0.659	0.640	0.625	0.612
1.7	0.953	0.912	0.844	0.791	0.749	0.715	0.688	0.666	0.649	0.635	0.623
1.8	0.953	0.913	0.845	0.793	0.752	0.720	0.694	0.674	0.657	0.644	0.634
1.9	0.953	0.913	0.847	0.796	0.756	0.725	0.701	0.681	0.666	0.654	0.644
2.0	0.954	0.913	0.848	0.798	0.760	0.730	0.707	0.688	0.674	0.662	0.653
2.1	0.954	0.914	0.849	0.801	0.763	0.735	0.712	0.695	0.682	0.671	0.663
2.2	0.954	0.914	0.851	0.803	0.767	0.739	0.718	0.702	0.689	0.679	0.672
2.3	0.954	0.915	0.852	0.805	0.770	0.744	0.724	0.708	0.696	0.687	0.680
2.4	0.954	0.915	0.853	0.808	0.774	0.748	0.729	0.714	0.703	0.695	0.688
2.5	0.954	0.915	0.854	0.810	0.777	0.753	0.734	0.720	0.710	0.702	0.696
2.6	0.954	0.916	0.856	0.812	0.780	0.757	0.739	0.726	0.717	0.709	0.704
2.7	0.954	0.916	0.857	0.814	0.783	0.761	0.744	0.732	0.723	0.716	0.711
2.8	0.954	0.916	0.858	0.817	0.787	0.765	0.749	0.737	0.729	0.723	0.718
2.9	0.955	0.917	0.859	0.819	0.790	0.769	0.754	0.743	0.735	0.729	0.725
3.0	0.955	0.917	0.860	0.821	0.793	0.773	0.758	0.748	0.740	0.735	0.731

Table B.23 – (continued)

β \ γ	1.2	1.4	1.6	1.8	2.0	2.3	2.6	2.9	3.2	3.6	4.0
0.1	0.325	0.274	0.233	0.199	0.172	0.140	0.116	0.099	0.086	0.074	0.065
0.2	0.348	0.301	0.263	0.232	0.207	0.179	0.157	0.142	0.130	0.119	0.112
0.3	0.371	0.327	0.292	0.264	0.242	0.216	0.197	0.183	0.174	0.164	0.158
0.4	0.392	0.352	0.321	0.295	0.275	0.252	0.236	0.224	0.215	0.208	0.203
0.5	0.413	0.376	0.348	0.325	0.307	0.287	0.272	0.262	0.255	0.249	0.245
0.6	0.433	0.400	0.374	0.353	0.337	0.320	0.307	0.299	0.293	0.288	0.285
0.7	0.453	0.422	0.398	0.380	0.366	0.351	0.341	0.334	0.329	0.325	0.322
0.8	0.471	0.443	0.422	0.406	0.394	0.380	0.372	0.366	0.362	0.359	0.357
0.9	0.489	0.463	0.444	0.430	0.419	0.408	0.401	0.396	0.393	0.391	0.389
1.0	0.506	0.483	0.466	0.453	0.444	0.434	0.428	0.425	0.422	0.420	0.418
1.1	0.522	0.501	0.486	0.475	0.467	0.459	0.454	0.451	0.449	0.447	0.446
1.2	0.538	0.519	0.506	0.496	0.489	0.482	0.478	0.475	0.474	0.472	0.471
1.3	0.553	0.536	0.524	0.515	0.509	0.504	0.500	0.498	0.497	0.495	0.494
1.4	0.567	0.552	0.541	0.534	0.529	0.524	0.521	0.519	0.518	0.517	0.515
1.5	0.581	0.567	0.558	0.551	0.547	0.543	0.540	0.539	0.538	0.536	0.535
1.6	0.594	0.582	0.573	0.568	0.564	0.560	0.558	0.557	0.556	0.555	0.553
1.7	0.606	0.595	0.588	0.583	0.580	0.577	0.575	0.574	0.573	0.572	0.570
1.8	0.618	0.608	0.602	0.598	0.595	0.593	0.591	0.590	0.589	0.587	0.586
1.9	0.630	0.621	0.615	0.612	0.609	0.607	0.606	0.605	0.604	0.602	0.600
2.0	0.641	0.633	0.628	0.625	0.622	0.621	0.619	0.618	0.617	0.616	0.614
2.1	0.651	0.644	0.640	0.637	0.635	0.633	0.632	0.631	0.630	0.629	0.627
2.2	0.661	0.655	0.651	0.648	0.647	0.645	0.644	0.643	0.642	0.641	0.639
2.3	0.671	0.665	0.661	0.659	0.658	0.657	0.656	0.655	0.654	0.652	0.650
2.4	0.680	0.674	0.671	0.670	0.668	0.667	0.666	0.665	0.664	0.663	0.661
2.5	0.688	0.684	0.681	0.679	0.678	0.677	0.676	0.675	0.674	0.673	0.671
2.6	0.697	0.692	0.690	0.689	0.688	0.687	0.686	0.685	0.684	0.682	0.680
2.7	0.705	0.701	0.699	0.698	0.697	0.696	0.695	0.694	0.693	0.691	0.689
2.8	0.712	0.709	0.707	0.706	0.705	0.704	0.703	0.702	0.701	0.699	0.697
2.9	0.719	0.716	0.715	0.714	0.713	0.712	0.711	0.710	0.709	0.707	0.705
3.0	0.726	0.724	0.722	0.721	0.721	0.720	0.719	0.718	0.717	0.715	0.713

Table B.24 – Heat exchangers – M and N type – 5 passages of external fluid – Fig. A.22 –Factor ψ

β	γ 0.05	0.10	0.20	0.30	0.40	0.50	0.60	0.70	0.80	0.90	1.00
0.1	0.951	0.905	0.820	0.744	0.676	0.614	0.559	0.509	0.464	0.424	0.387
0.2	0.951	0.906	0.822	0.747	0.681	0.622	0.569	0.521	0.479	0.440	0.406
0.3	0.952	0.906	0.824	0.751	0.686	0.629	0.578	0.533	0.492	0.457	0.424
0.4	0.952	0.907	0.825	0.754	0.691	0.636	0.587	0.544	0.506	0.472	0.442
0.5	0.952	0.907	0.827	0.757	0.696	0.643	0.596	0.555	0.519	0.487	0.459
0.6	0.952	0.907	0.828	0.760	0.701	0.650	0.605	0.566	0.532	0.502	0.476
0.7	0.952	0.908	0.830	0.763	0.706	0.656	0.614	0.577	0.544	0.516	0.492
0.8	0.952	0.908	0.831	0.766	0.710	0.663	0.622	0.587	0.556	0.530	0.507
0.9	0.952	0.909	0.833	0.769	0.715	0.669	0.630	0.597	0.568	0.543	0.522
1.0	0.952	0.909	0.834	0.772	0.719	0.675	0.638	0.606	0.579	0.556	0.536
1.1	0.953	0.910	0.836	0.775	0.724	0.681	0.646	0.616	0.590	0.568	0.550
1.2	0.953	0.910	0.837	0.777	0.728	0.687	0.653	0.625	0.601	0.580	0.563
1.3	0.953	0.910	0.839	0.780	0.732	0.693	0.660	0.633	0.611	0.592	0.576
1.4	0.953	0.911	0.840	0.783	0.736	0.699	0.668	0.642	0.621	0.603	0.588
1.5	0.953	0.911	0.841	0.785	0.741	0.704	0.674	0.650	0.630	0.614	0.600
1.6	0.953	0.912	0.843	0.788	0.744	0.709	0.681	0.658	0.639	0.624	0.612
1.7	0.953	0.912	0.844	0.791	0.748	0.715	0.688	0.666	0.648	0.634	0.622
1.8	0.953	0.913	0.845	0.793	0.752	0.720	0.694	0.673	0.657	0.644	0.633
1.9	0.953	0.913	0.847	0.796	0.756	0.725	0.700	0.681	0.665	0.653	0.643
2.0	0.954	0.913	0.848	0.798	0.760	0.730	0.706	0.688	0.673	0.662	0.653
2.1	0.954	0.914	0.849	0.801	0.763	0.734	0.712	0.695	0.681	0.670	0.662
2.2	0.954	0.914	0.851	0.803	0.767	0.739	0.718	0.701	0.688	0.679	0.671
2.3	0.954	0.915	0.852	0.805	0.770	0.744	0.723	0.708	0.696	0.686	0.679
2.4	0.954	0.915	0.853	0.808	0.774	0.748	0.729	0.714	0.703	0.694	0.688
2.5	0.954	0.915	0.854	0.810	0.777	0.752	0.734	0.720	0.709	0.702	0.696
2.6	0.954	0.916	0.856	0.812	0.780	0.756	0.739	0.726	0.716	0.709	0.703
2.7	0.954	0.916	0.857	0.814	0.783	0.761	0.744	0.732	0.722	0.716	0.710
2.8	0.954	0.916	0.858	0.816	0.786	0.765	0.749	0.737	0.728	0.722	0.717
2.9	0.955	0.917	0.859	0.819	0.789	0.769	0.753	0.742	0.734	0.728	0.724
3.0	0.955	0.917	0.860	0.821	0.792	0.772	0.758	0.748	0.740	0.735	0.731

Table B.24 – (continued)

	γ										
β	1.2	1.4	1.6	1.8	2.0	2.3	2.6	2.9	3.2	3.6	4.0
0.1	0.325	0.274	0.233	0.199	0.171	0.139	0.115	0.097	0.084	0.071	0.063
0.2	0.348	0.301	0.262	0.231	0.206	0.177	0.155	0.139	0.127	0.115	0.107
0.3	0.370	0.326	0.291	0.263	0.240	0.214	0.194	0.180	0.169	0.159	0.152
0.4	0.391	0.351	0.319	0.294	0.273	0.249	0.232	0.220	0.211	0.202	0.196
0.5	0.412	0.375	0.346	0.323	0.304	0.284	0.269	0.258	0.251	0.243	0.239
0.6	0.432	0.398	0.372	0.351	0.335	0.317	0.304	0.295	0.288	0.283	0.279
0.7	0.451	0.420	0.396	0.378	0.364	0.348	0.337	0.330	0.324	0.320	0.317
0.8	0.470	0.442	0.420	0.404	0.391	0.377	0.368	0.362	0.358	0.355	0.352
0.9	0.488	0.462	0.443	0.428	0.417	0.405	0.398	0.393	0.390	0.387	0.385
1.0	0.505	0.481	0.464	0.451	0.442	0.432	0.426	0.422	0.419	0.417	0.416
1.1	0.521	0.500	0.484	0.473	0.465	0.457	0.452	0.448	0.446	0.445	0.444
1.2	0.537	0.518	0.504	0.494	0.487	0.480	0.476	0.473	0.472	0.471	0.470
1.3	0.552	0.534	0.522	0.514	0.508	0.502	0.498	0.497	0.495	0.494	0.494
1.4	0.566	0.550	0.540	0.532	0.527	0.522	0.520	0.518	0.517	0.516	0.516
1.5	0.580	0.566	0.556	0.550	0.545	0.541	0.539	0.538	0.537	0.537	0.536
1.6	0.593	0.580	0.572	0.566	0.563	0.559	0.558	0.557	0.556	0.555	0.555
1.7	0.605	0.594	0.587	0.582	0.579	0.576	0.575	0.574	0.573	0.573	0.572
1.8	0.617	0.607	0.601	0.597	0.594	0.592	0.591	0.590	0.590	0.589	0.588
1.9	0.629	0.620	0.614	0.611	0.608	0.607	0.606	0.605	0.605	0.604	0.603
2.0	0.640	0.632	0.627	0.624	0.622	0.620	0.620	0.619	0.619	0.618	0.617
2.1	0.650	0.643	0.639	0.636	0.635	0.633	0.633	0.632	0.632	0.631	0.630
2.2	0.660	0.654	0.650	0.648	0.646	0.645	0.645	0.644	0.644	0.643	0.642
2.3	0.670	0.664	0.661	0.659	0.658	0.657	0.656	0.656	0.655	0.654	0.653
2.4	0.679	0.674	0.671	0.669	0.668	0.668	0.667	0.667	0.666	0.665	0.664
2.5	0.688	0.683	0.681	0.679	0.678	0.678	0.677	0.677	0.676	0.675	0.674
2.6	0.696	0.692	0.690	0.689	0.688	0.687	0.687	0.686	0.686	0.685	0.684
2.7	0.704	0.700	0.699	0.697	0.697	0.696	0.696	0.695	0.695	0.694	0.692
2.8	0.712	0.708	0.707	0.706	0.705	0.705	0.704	0.704	0.703	0.702	0.701
2.9	0.719	0.716	0.715	0.714	0.713	0.713	0.713	0.712	0.711	0.710	0.709
3.0	0.726	0.723	0.722	0.721	0.721	0.721	0.720	0.720	0.719	0.718	0.717

B.4 Coils

B.4.1 Coils with Fluids in Parallel Flow (Fig. 2.8)

Table B.25 – Coils – parallel flow – 2 sections – Fig. A.23 – Corrective factor φ_p

	γ			
β	0.70	0.80	0.90	1.00
0.7	....	....	....	0.990
0.8	....	....	0.991	0.990
0.9	....	....	0.992	0.990
1.0	....	0.993	0.992	0.991
1.1	....	0.993	0.992	0.991
1.2	....	0.993	0.992	0.992
1.3	0.995	0.994	0.993	0.992
1.4	0.995	0.994	0.993	0.993
1.5	0.995	0.994	0.994	0.994
1.6	0.995	0.994	0.994	....
1.7	0.995	0.995	....	....
1.8	0.995	0.995	....	....

If $\gamma < 0.7$ use factor ψ_p without corrective factor

Table B.25 – (continued)

β	γ 1.2	1.4	1.6	1.8	2.0	2.3	2.6	2.9	3.2	3.6	4.0
0.3	....	....	....	0.977	0.974	0.972	....	....	....	....	....
0.4	....	0.983	0.980	0.978	0.976	0.977	....	....	....	....	....
0.5	0.987	0.983	0.981	0.980	0.980	....	....	....	....	....	1.024
0.6	0.987	0.984	0.983	0.983	0.984	....	....	....	....	1.023	1.035
0.7	0.987	0.985	0.985	0.986	....	....	....	....	1.019	1.032	1.044
0.8	0.988	0.987	0.987	....	....	....	....	1.017	1.027	1.040	1.053
0.9	0.989	0.988	0.990	....	....	....	1.014	1.024	1.034	1.047	1.060
1.0	0.990	0.990	....	....	....	....	1.019	1.029	1.040	1.053	1.066
1.1	0.991	....	....	....	....	1.013	1.024	1.035	1.045	1.059	1.071
1.2	0.992	....	....	....	....	1.017	1.028	1.039	1.050	1.063	1.075
1.3	0.993	....	....	....	1.011	1.021	1.032	1.043	1.054	1.067	1.078
1.4	....	....	....	1.007	1.014	1.024	1.035	1.046	1.057	1.070	1.081
1.5	....	....	....	1.010	1.016	1.027	1.038	1.049	1.060	1.072	1.083
1.6	....	....	....	1.012	1.019	1.030	1.041	1.052	1.062	1.074	1.085
1.8	....	....	1.009	1.016	1.023	1.034	1.045	1.056	1.066	1.077	1.087
2.0	....	1.006	1.012	1.019	1.027	1.038	1.049	1.059	1.068	1.079	1.088
2.2	....	1.009	1.015	1.022	1.030	1.041	1.051	1.061	1.069	1.079	1.088
2.6	1.007	1.013	1.019	1.027	1.034	1.044	1.054	1.062	1.070	1.079	1.086
3.0	1.009	1.016	1.023	1.029	1.036	1.046	1.054	1.062	1.069	1.076	1.083

Table B.26 – Coils – parallel flow – 3 sections – Fig. A.24 – Corrective factor φ_p

β	γ 1.4	1.6	1.8	2.0	2.3	2.6	2.9	3.2	3.6	4.0
1.2	....	....	....	....	....	....	....	....	0.990	0.987
1.3	....	....	....	....	....	....	....	0.992	0.989	0.986
1.4	....	....	....	....	....	....	0.992	0.991	0.987	0.984
1.5	....	....	....	....	....	....	0.992	0.989	0.986	0.982
1.6	....	....	....	....	....	0.993	0.991	0.988	0.985	0.980
1.8	....	....	....	....	0.994	0.992	0.989	0.986	0.982	0.977
2.0	....	....	....	0.995	0.993	0.990	0.987	0.984	0.979	0.975
2.2	....	....	0.995	0.994	0.992	0.989	0.986	0.982	0.977	0.973
2.6	....	0.996	0.994	0.993	0.990	0.987	0.983	0.979	0.974	0.970
3.0	0.996	0.995	0.993	0.992	0.988	0.985	0.981	0.977	0.973	0.969

If $\beta < 1.2$ or $\gamma < 1.4$ use factor ψ_p without corrective factor

B.4.2 Coils with Fluids in Counter Flow (Fig. 2.9)

Table B.27 – Coils – counter flow – 2 sections – Fig. A.26 – Corrective factor φ_c

β	γ 0.5	0.6	0.7	0.8	0.9	1.0
0.6	....	....	....	....	1.012	1.016
0.8	....	....	....	1.010	1.014	1.019
1.0	....	....	1.008	1.011	1.015	1.020
1.2	....	1.006	1.009	1.012	1.016	1.020
1.4	....	1.006	1.009	1.012	1.016	1.020
1.6	....	1.006	1.009	1.012	1.015	1.019
1.8	1.004	1.006	1.009	1.012	1.015	1.018
2.0	1.004	1.006	1.009	1.011	1.014	1.016
2.2	1.004	1.006	1.008	1.011	1.013	1.015
2.4	1.004	1.006	1.008	1.010	1.012	1.014
2.6	1.004	1.006	1.008	1.009	1.011	1.012
2.8	1.004	1.006	1.007	1.009	1.010	1.011
3.0	1.004	1.005	1.007	1.008	1.009	1.010

If $\gamma < 0.5$ use factor ψ_c without corrective factor

Table B.27 – (continued)

β \ γ	1.2	1.4	1.6	1.8	2.0	2.3	2.6	2.9	3.2	3.6	4.0
0.2	....	....	....	1.042	1.057	1.087	1.127	1.179	1.246	1.363	1.520
0.3	....	1.027	1.040	1.056	1.076	1.115	1.167	1.233	1.316	1.460	1.651
0.4	1.021	1.033	1.048	1.067	1.090	1.134	1.191	1.264	1.354	1.507	1.705
0.5	1.024	1.037	1.054	1.074	1.099	1.145	1.204	1.277	1.366	1.514	1.701
0.6	1.027	1.040	1.057	1.078	1.104	1.150	1.207	1.276	1.360	1.494	1.659
0.7	1.028	1.042	1.060	1.080	1.105	1.149	1.202	1.266	1.340	1.456	1.595
0.8	1.030	1.043	1.060	1.080	1.104	1.145	1.193	1.249	1.312	1.410	1.522
0.9	1.030	1.044	1.060	1.079	1.100	1.137	1.180	1.228	1.281	1.360	1.447
1.0	1.030	1.044	1.059	1.076	1.096	1.129	1.165	1.205	1.248	1.310	1.376
1.1	1.030	1.043	1.057	1.073	1.090	1.119	1.149	1.182	1.216	1.264	1.312
1.2	1.030	1.042	1.055	1.069	1.084	1.108	1.134	1.160	1.186	1.222	1.257
1.3	1.029	1.040	1.052	1.064	1.078	1.098	1.118	1.139	1.159	1.185	1.209
1.4	1.029	1.038	1.049	1.060	1.071	1.088	1.104	1.120	1.134	1.153	1.169
1.5	1.028	1.036	1.046	1.055	1.065	1.078	1.091	1.102	1.113	1.126	1.136
1.6	1.026	1.034	1.043	1.051	1.058	1.069	1.079	1.087	1.095	1.103	1.109
1.7	1.025	1.032	1.039	1.046	1.052	1.061	1.068	1.074	1.079	1.084	1.087
1.8	1.024	1.030	1.036	1.042	1.047	1.053	1.058	1.062	1.065	1.068	1.070
1.9	1.023	1.028	1.033	1.038	1.042	1.047	1.050	1.053	1.054	1.055	1.056
2.0	1.022	1.026	1.031	1.034	1.037	1.041	1.043	1.044	1.045	1.045	1.044
2.2	1.019	1.023	1.025	1.028	1.029	1.030	1.031	1.031	1.030	1.029	1.028
2.4	1.017	1.019	1.021	1.022	1.023	1.023	1.022	1.021	1.021	1.019	1.018
2.6	1.015	1.016	1.017	1.017	1.017	1.017	1.016	1.015	1.014	1.013	1.011
2.8	1.013	1.013	1.014	1.014	1.013	1,012	1,011	1.010	1.009	1.008	1.007
3.0	1.011	1.011	1.011	1.011	1.010	1.009	1.008	1.007	1.006	1.005	-----

Table B.28 – Coils – counter flow – 3 sections – Fig. A.27 – Corrective factor φ_c

β	γ 1.2	1.4	1.6	1.8	2.0	2.3	2.6	2.9	3.2	3.6	4.0
0.3	....	....	....	....	....	....	....	1.100	1.135	1.194	1.270
0.4	....	....	....	....	1.040	1.059	1.085	1.116	1.155	1.221	1.306
0.5	....	....	....	1.033	1.044	1.065	1.092	1.125	1.166	1.233	1.317
0.6	....	....	1.026	1.035	1.047	1.068	1.095	1.128	1.167	1.231	1.311
0.7	....	1.019	1.027	1.037	1.048	1.069	1.095	1.126	1.162	1.221	1.291
0.8	....	1.020	1.028	1.037	1.048	1.068	1.092	1.120	1.153	1.203	1.263
0.9	1.014	1.020	1.028	1.037	1.047	1.066	1.087	1.112	1.140	1.182	1.230
1.0	1.014	1.020	1.027	1.036	1.046	1.062	1.081	1.102	1.126	1.160	1.197
1.1	1.014	1.020	1.027	1.035	1.043	1.058	1.074	1.092	1.111	1.138	1.166
1.2	1.014	1.019	1.026	1.033	1.041	1.054	1.067	1.081	1.096	1.117	1.137
1.3	1.014	1.019	1.025	1.031	1.038	1.049	1.060	1.071	1.083	1.097	1.112
1.4	1.013	1.018	1.024	1.029	1.035	1.044	1.053	1.062	1.070	1.080	1.090
1.5	1.013	1.017	1.022	1.027	1.032	1.039	1.046	1.053	1.059	1.066	1.072
1.6	1.013	1.017	1.021	1.025	1.029	1.035	1.040	1.045	1.049	1.053	1.057
1.7	1.012	1.016	1.019	1.023	1.026	1.031	1.035	1.038	1.040	1.041	1.045
1.8	1.011	1.015	1.018	1.021	1.024	1.027	1.030	1.032	1.033	1.034	1.035
1.9	1.011	1.014	1.016	1.019	1.021	1.024	1.025	1.027	1.027	1.027	1.027
2.0	1.010	1.013	1.015	1.017	1.019	1.020	1.022	1.022	1.022	1.022	1.021
2.2	1.009	1.011	1.013	1.014	1.015	1.015	1.015	1.015	1.015	1.014	1.013
2.6	1.007	1.008	1.008	1.009	1.009	1.008	1.008	1.007	....	....	....
3.0	1.005	1.006	1.005	1.005	1.005	....	....	....	....	....	....

If $\beta < 0.3$ or $\gamma < 1.2$ use factor ψ_c without corrective factor

Table B.29 – Coils – counter flow – 4 sections – Fig. A.28 – Corrective factor φ_c

	γ									
β	1.4	1.6	1.8	2.0	2.3	2.6	2.9	3.2	3.6	4.0
0.4	…….	…….	…….	…….	…….	…….	…….	…….	…….	1.171
0.5	…….	…….	…….	…….	…….	…….	1.071	1.095	1.133	1.180
0.6	…….	…….	…….	…….	1.039	1.055	1.074	1.096	1.133	1.179
0.7	…….	…….	…….	1.028	1.040	1.055	1.073	1.095	1.129	1.170
0.8	…….	…….	1.021	1.028	1.040	1.054	1.070	1.090	1.120	1.156
0.9	…….	…….	1.021	1.028	1.038	1.051	1.066	1.083	1.109	1.138
1.0	…….	1.016	1.021	1.027	1.037	1.048	1.061	1.075	1.096	1.120
1.1	…….	1.016	1.020	1.025	1.034	1.044	1.055	1.067	1.083	1.101
1.2	…….	1.015	1.019	1.024	1.032	1.040	1.049	1.058	1.071	1.084
1.3	…….	1.014	1.018	1.022	1.029	1.036	1.043	1.050	1.059	1.068
1.4	1.011	1.014	1.017	1.021	1.026	1.032	1.037	1.042	1.049	1.055
1.5	1.010	1.013	1.016	1.019	1.023	1.028	1.032	1.036	1.040	1.044
1.6	1.010	1.012	1.015	1.017	1.021	1.024	1.027	1.030	1.032	1.034
1.7	1.009	1.011	1.014	1.016	1.018	1.021	1.023	1.024	1.026	1.027
1.8	1.009	1.011	1.012	1.014	1.016	1.018	1.019	1.020	1.021	1.021
1.9	1.008	1.010	1.011	1.013	1.014	1.015	1.016	1.016	1.016	1.016
2.0	1.008	1.009	1.010	1.011	1.012	1.013	1.013	1.013	1.013	1.012
2.2	…….	1.007	1.008	1.009	1.009	1.009	1.009	1.008	…….	…….
2.4	…….	…….	1.007	1.007	…….	…….	…….	…….	…….	…….

If $\beta < 0.4$ or $\beta > 2.4$ or $\gamma < 1.4$ use factor ψ_c without corrective factor

Table B.30 – Coils – counter flow – 6 sections – Fig. A.29 – Corrective factor φ_c

	γ			
β	2.9	3.2	3.6	4.0
0.7	....	....	....	1.077
0.8	....	....	1.055	1.072
0.9	....	1.038	1.050	1.065
1.0	....	1.035	1.045	1.056
1.1	1.025	1.031	1.039	1.048
1.2	1.023	1.027	1.033	1.040
1.3	1.020	1.023	1.028	1.033
1.4	1.017	1.020	1.023	1.026
1.5	....	1.017	1.019	1.021
1.6	....	....	1.015	1.016

If $\beta < 0.7$ or $\beta > 1.6$ or $\gamma < 2.9$ use factor ψ_c without corrective factor

B.5 Tube Banks with Several Passages of External Fluid

B.5.1 Tube Banks with Fluids in Parallel Flow (Fig. 2.10)

Table B.31 – Tube banks – parallel flow – 2 passages of external fluid Fig. A.32 – Corrective factor φ_p

	γ				
β	0.6	0.7	0.8	0.9	1.0
0.5	....	....	....	....	0.989
0.6	....	....	....	0.991	0.989
0.7	....	....	....	0.991	0.989
0.8	....	....	0.992	0.990	0.989
0.9	....	....	0.992	0.991	0.989
1.0	....	0.994	0.992	0.991	0.989
1.1	....	0.994	0.992	0.991	0.990
1.2	....	0.994	0.993	0.991	0.991
1.3	....	0.994	0.993	0.992	0.991
1.4	....	0.994	0.993	0.992	0.992
1.5	0.995	0.994	0.993	0.993	0.993
1.6	0.996	0.995	0.994	0.993	0.994
1.7	0.996	0.995	0.994	0.994	0.994
1.8	0.996	0.995	0.995	0.995	....
1.9	0.996	0.995	0.995	0.995	....
2.0	0.996	0.995	0.995	....	....
2.1	0.996	0.996	....	....	....
2.2	0.996	0.996	....	....	....

If $\beta < 0.5$ or $\beta > 2.2$ or $\gamma < 0.5$ use factor ψ_p without corrective factor

Table B.31 – (continued)

β	γ 1.2	1.4	1.6	1.8	2.0	2.3	2.6	2.9	3.2	3.6	4.0
0.2	....	....	....	....	0.967	0.960	0.956	0.954	0.955	....	....
0.3	....	0.981	0.976	0.971	0.967	0.963	0.962	0.964	....	....	....
0.4	0.985	0.980	0.976	0.972	0.970	0.969	0.971	0.976	....	....	....
0.5	0.984	0.980	0.977	0.974	0.974	0.975	0.980	....	....	....	1.031
0.6	0.984	0.981	0.979	0.978	0.978	0.982	....	....	....	1.028	1.047
0.7	0.985	0.982	0.981	0.981	0.983	....	....	....	1.022	1.041	1.060
0.8	0.986	0.984	0.984	0.985	0.988	....	....	1.018	1.032	1.052	1.072
0.9	0.987	0.986	0.986	0.989	....	....	1.013	1.026	1.041	1.061	1.081
1.0	0.988	0.988	0.989	....	....	....	1.020	1.034	1.048	1.069	1.089
1.1	0.989	0.990	....	....	....	1.013	1.026	1.040	1.055	1.075	1.095
1.2	0.990	0.992	....	....	....	1.018	1.031	1.046	1.061	1.081	1.100
1.3	0.991	....	....	....	1.010	1.022	1.036	1.050	1.065	1.085	1.104
1.4	0.993	....	....	....	1.013	1.026	1.040	1.055	1.069	1.089	1.107
1.5	....	....	....	1.009	1.016	1.029	1.044	1.058	1.073	1.092	1.110
1.6	....	....	....	1.011	1.019	1.033	1.047	1.061	1.075	1.094	1.111
1.7	....	....	1.007	1.014	1.022	1.035	1.049	1.064	1.077	1.095	1.112
1.8	....	....	1.009	1.016	1.024	1.038	1.052	1.066	1.079	1.096	1.113
1.9	....	....	1.011	1.018	1.027	1.040	1.054	1.067	1.080	1.097	1.113
2.0	....	1.006	1.012	1.020	1.029	1.042	1.055	1.069	1.081	1.098	1.112
2.1	....	1.007	1.014	1.022	1.030	1.043	1.057	1.070	1.082	1.098	1.112
2.2	....	1.008	1.015	1.023	1.032	1.045	1.058	1.070	1.082	1.097	1.111
2.3	....	1.010	1.017	1.025	1.033	1.046	1.059	1.071	1.083	1.097	1.110
2.4	1.004	1.011	1.018	1.026	1.034	1.047	1.059	1.071	1.083	1.096	1.109
2.5	1.005	1.012	1.019	1.027	1.035	1.048	1.060	1.072	1.082	1.096	1.108
2.6	1.006	1.013	1.020	1.028	1.036	1.049	1.060	1.072	1.082	1.095	1.107
2.8	1.008	1.015	1.022	1.030	1.038	1.050	1.061	1.071	1.081	1.093	1.104
3.0	1.009	1.016	1.024	1.031	1.039	1.050	1.061	1.071	1.080	1.091	1.101

Table B.32 – Tube banks – parallel flow – 3 passages of external fluid – Fig. A.33 – Corrective factor φ_p

	γ									
β	1.4	1.6	1.8	2.0	2.3	2.6	2.9	3.2	3.6	4.0
1.0	....	....	....	....	....	....	....	....	....	0.990
1.1	....	....	....	....	....	....	....	....	0.991	0.988
1.2	....	....	....	....	....	....	....	....	0.989	0.986
1.3	....	....	....	....	....	....	....	0.991	0.988	0.983
1.4	....	....	....	....	....	....	0. 992	0.990	0.986	0.981
1.5	....	....	....	....	....	0.993	0.991	0.989	0.984	0.979
1.6	....	....	....	....	....	0.993	0.990	0.987	0.982	0.977
1.7	....	....	....	....	0.994	0.992	0.989	0.986	0.981	0.975
1.8	....	....	....	....	0.993	0.991	0.988	0.985	0.979	0.973
1.9	....	....	....	0.995	0.993	0.991	0.987	0.984	0.978	0.971
2.0	....	....	....	0.994	0.993	0.990	0.986	0.982	0.976	0.970
2.1	....	....	0.995	0.994	0.992	0.9 89	0.985	0.981	0.975	0.968
2.2	....	....	0.995	0.994	0.992	0.9 88	0.984	0.980	0.974	0.967
2.3	....	....	0.995	0.994	0.991	0.9 88	0.984	0.979	0.973	0.966
2.4	....	0.996	0.995	0.993	0.990	0.987	0.983	0.978	0.972	0.965
2.6	....	0.996	0.994	0.993	0.989	0.986	0.981	0.977	0.970	0.964
2.8	0.996	0.995	0.994	0.992	0.988	0.984	0.980	0.975	0.969	0.963
3.0	0.996	0.995	0.993	0.991	0.988	0.983	0.979	0.974	0.968	0.963

If β< 1.0 or γ< 1.4 use factor. ψ_p without the corrective factor

B.5.2 Tube Banks with Fluids in Counter Flow (Fig. 2.11)

Table B.33 – Tube banks – counter flow – 2 passages of external fluid Fig. A.34 – Corrective factor φ_c

	γ					
β	0.5	0.6	0.7	0.8	0.9	1.0
0.5	....	....	....	....	....	1.013
0.6	....	....	....	....	1.011	1.014
0.7	....	....	....	....	1.012	1.016
0.8	....	....	....	1.009	1.013	1.016
0.9	....	....	....	1.010	1.013	1.017
1.0	....	....	1.007	1.010	1.014	1.017
1.1	....	....	1.008	1.011	1.014	1.018
1.2	....	....	1.008	1.011	1.014	1.018
1.3	....	1.006	1.008	1.011	1.014	1.018
1.4	....	1.006	1.008	1.011	1.014	1.017
1.5	....	1.006	1.008	1.011	1.014	1.017
1.6	....	1.006	1.008	1.011	1.014	1.016
1.7	....	1.006	1.008	1.011	1.013	1.016
1.8	....	1.006	1.008	1.010	1.013	1.015
1.9	1.004	1.006	1.008	1.010	1.013	1.015
2.0	1.004	1.006	1.008	1.010	1.012	1.014
2.2	1.004	1.006	1.008	1.010	1.011	1.013
2.4	1.004	1.006	1.007	1.009	1.010	1.012
2.6	1.004	1.005	1.007	1.008	1.010	1.011
2.8	1.004	1.005	1.007	1.008	1.009	1.009
3.0	1.004	1.005	1.006	1.007	1.008	1.008

If $\beta < 0.5$ or $\gamma < 0.5$ use factor ψ_c without corrective factor

Table B.33 – (continued)

	γ										
β	1.2	1.4	1.6	1.8	2.0	2.3	2.6	2.9	3.2	3.6	4.0
0.2	....	....	....	....	....	....	1.088	1.118	1.154	1.212	....
0.3	....	....	1.032	1.044	1.058	1.084	1.116	1.155	1.200	1.273	1.361
0.4	1.018	1.027	1.039	1.053	1.069	1.098	1.134	1.177	1.227	1.306	1.400
0.5	1.021	1.031	1.043	1.058	1.076	1.107	1.144	1.188	1.238	1.316	1.407
0.6	1.023	1.034	1.047	1.062	1.080	1.111	1.147	1.189	1.236	1.308	1.390
0.7	1.024	1.035	1.048	1.064	1.081	1.110	1.144	1.182	1.225	1.288	1.358
0.8	1.025	1.036	1.049	1.064	1.080	1.107	1.138	1.172	1.208	1.261	1.318
0.9	1.026	1.037	1.049	1.063	1.078	1.102	1.129	1.158	1.188	1.230	1.274
1.0	1.026	1.036	1.048	1.060	1.074	1.096	1.118	1.142	1.166	1.199	1.232
1.1	1.026	1.036	1.046	1.058	1.070	1.088	1.107	1.126	1.144	1.169	1.192
1.2	1.026	1.035	1.044	1.054	1.065	1.080	1.095	1.110	1.124	1.141	1.157
1.3	1.025	1.033	1.042	1.051	1.060	1.072	1.084	1.095	1.105	1.117	1.126
1.4	1.024	1.032	1.040	1.047	1.054	1.065	1.074	1.082	1.088	1.096	1.101
1.5	1.023	1.030	1.037	1.043	1.049	1.057	1.064	1.069	1.074	1.077	1.080
1.6	1.023	1.029	1.034	1.040	1.044	1.051	1.055	1.059	1.061	1.062	1.062
1.7	1.022	1.027	1.032	1.036	1.040	1.044	1.047	1.049	1.050	1.050	1.049
1.8	1.020	1.025	1.029	1.033	1.036	1.039	1.040	1.041	1.041	1.040	1.038
1.9	1.019	1.023	1.027	1.029	1.032	1.033	1.034	1.034	1.033	1.031	1.029
2.0	1.018	1.022	1.024	1.026	1.028	1.029	1.029	1.028	1.027	1.025	1.022
2.1	1.017	1.020	1.022	1.024	1.025	1.025	1.024	1.023	1.022	1.019	1.017
2.2	1.016	1.019	1.020	1.021	1.022	1.021	1.020	1.019	1.017	1.015	1.013
2.3	1.015	1.017	1.018	1.019	1.019	1.018	1.017	1.016	1.014	1.012	1.010
2.4	1.014	1.016	1.016	1.017	1.016	1.016	1.014	1.013	1,011	1.009	1.008
2.5	1.013	1.014	1.015	1.015	1.014	1.013	1.012	1.010	1.009	1.007	....
2.6	1.012	1.013	1.013	1.013	1.012	1.011	1.010	1.008	1.007	....	....
2.8	1.010	1.011	1.011	1.010	1.009	1.008	1.007	1.006	....	....	....
3.0	1.009	1.009	1.008	1.008	1.007	1.006	....	....	....	....	....

Table B.34 – Tube banks – counter flow – 3 passages of external fluid – Fig. A.35 – Corrective factor φ_c

β	γ 0.9	1.1	1.4	1.7	2.0	2.3	2.6	2.9	3.2	3.6	4.0
0.3	....	....	....	....	....	....	....	....	1.123	1.174	1.238
0.4	....	....	....	....	....	1.056	1.079	1.107	1.141	1.198	1.269
0.5	....	....	....	1.027	1.042	1.061	1.085	1.115	1.151	1.208	1.279
0.6	....	....	....	1.029	1.044	1.064	1.088	1.118	1.152	1.207	1.273
0.7	....	....	1.018	1.030	1.046	1.065	1.088	1.116	1.148	1.198	1.256
0.8	....	....	1.019	1.031	1.046	1.064	1.085	1.110	1.139	1.182	1.232
0.9	....	....	1.019	1.031	1.045	1.062	1.081	1.103	1.127	1.163	1.203
1.0	....	....	1.019	1.030	1.043	1.058	1.075	1.094	1.114	1.143	1.174
1.1	....	1.011	1.019	1.029	1.041	1.054	1.069	1.084	1.101	1.123	1.146
1.2	....	1.011	1.019	1.028	1.039	1.050	1.062	1.075	1.087	1.104	1.121
1.3	....	1.011	1.018	1.027	1.036	1.046	1.055	1.065	1.075	1.087	1.098
1.4	....	1.011	1.017	1.025	1.033	1.041	1.049	1.057	1.064	1.072	1.079
1.5	....	1.011	1.017	1.023	1.030	1.037	1.043	1.048	1.053	1.059	1.063
1.6	....	1.010	1.016	1.022	1.027	1.033	1.037	1.041	1.044	1.048	1.050
1.7	1.007	1.010	1.015	1.020	1.025	1.029	1.032	1.035	1.037	1.038	1.039
1.8	1.007	1.010	1.014	1.018	1.022	1.025	1.027	1.029	1.030	1.031	1.030
1.9	1.006	1.009	1.013	1.017	1.020	1.022	1.023	1.024	1.025	1.024	1.024
2.0	1.006	1.009	1.012	1.015	1.018	1.019	1.020	1.020	1.020	1.019	1.018
2.1	1.006	1.008	1.011	1.014	1.016	1.016	1.017	1.017	1.016	1.015	1.014
2.2	1.006	1.008	1.011	1.013	1.014	1.014	1.014	1.014	1.013	1.012	1.011
2.3	1.006	1.008	1.010	1.011	1.012	1.012	1.012	1.011	1.011	1.009	1.008
2.4	1.005	1.007	1.009	1.010	1.011	1.010	1.010	1.009	1.008	1.007	....
2.6	1.005	1.006	1.008	1.008	1.008	1.008	1.007	....	....	....	....
2.8	1.005	1.006	1.006	1.006	1.006	1.005	....	....	....	....	....
3.0	1.004	1.005	1.005	1.005	....	....	....	....	....	....	....

If $\beta < 0.3$ or $\gamma < 0.9$ use factor ψ_c without corrective factor

Table B.35 – Tube banks – counter flow – 4 passages of external fluid – Fig. A.36 – Corrective factor φ_c

	γ									
β	1.4	1.6	1.8	2.0	2.3	2.6	2.9	3.2	3.6	4.0
0.4	…….	…….	…….	…….	…….	…….	…….	…….	…….	1.170
0.5	…….	…….	…….	…….	…….	…….	1.072	1.094	1.132	1.179
0.6	…….	…….	…….	…….	1.039	1.055	1.074	1.096	1.133	1.178
0.7	…….	…….	…….	1.028	1.040	1.055	1.073	1.094	1.128	1.169
0.8	…….	…….	…….	1.028	1.040	1.054	1.070	1.090	1.120	1.155
0.9	…….	…….	1.021	1.028	1.038	1.051	1.066	1.083	1.108	1.138
1.0	…….	…….	1.021	1.027	1.037	1.048	1.061	1.075	1.096	1.119
1.1	…….	1.016	1.020	1.025	1.034	1.044	1.055	1.066	1.083	1.101
1.2	…….	1.015	1.019	1.024	1.032	1.040	1.049	1.058	1.071	1.084
1.3	…….	1.014	1.018	1.022	1.029	1.036	1.043	1.050	1.059	1.068
1.4	…….	1.014	1.017	1.021	1.026	1.032	1.037	1.042	1.049	1.055
1.5	1.010	1.013	1.016	1.019	1.023	1.028	1.032	1.036	1.040	1.044
1.6	1.010	1.012	1.015	1.017	1.021	1.024	1.027	1.030	1.032	1.034
1.7	1.009	1.011	1.014	1.016	1.018	1.021	1.023	1.024	1.026	1.027
1.8	1.009	1.011	1.012	1.014	1.016	1.018	1.019	1.020	1.021	1.021
1.9	1.008	1.010	1.011	1.013	1.014	1.015	1.016	1.016	1.016	1.016
2.0	1.008	1.009	1.010	1.011	1.012	1.013	1.013	1.013	1.013	1.012
2.1	…….	1.008	1.009	1.010	1.011	1.011	1.011	1.011	1.010	1.009
2.2	…….	1.007	1.008	1.009	1.009	1.009	1.009	1.009	…….	…….
2.3	…….	1.007	1.007	1.008	1.008	1.008	…….	…….	…….	…….

If $\beta < 0.4$ or $\beta > 2.3$ or $\gamma < 1.4$ use factor ψ_c without corrective factor

CPSIA information can be obtained at www.ICGtesting.com
Printed in the USA
LVOW100118181011

250934LV00009B/122/P

9 783642 133084